Quality and Legitimacy of Global Governance

Case Lessons from Forestry

Timothy Cadman

Fellow, Sustainable Business, University of Southern Queensland, Australia

First published 2011 by
PALGRAVE MACMILLAN

Palgrave Macmillan in the UK is an imprint of Macmillan Publishers Limited,
registered in England, company number 785998, of Houndmills, Basingstoke,
Hampshire RG21 6XS.

Palgrave Macmillan in the US is a division of St Martin's Press LLC,
175 Fifth Avenue, New York, NY 10010.

Palgrave Macmillan is the global academic imprint of the above companies
and has companies and representatives throughout the world.

Palgrave® and Macmillan® are registered trademarks in the United States,
the United Kingdom, Europe and other countries.

ISBN 978–0–230–24358–3 hardback

A catalogue record for this book is available from the British Library.

A catalog record for this book is available from the Library of Congress.

10   9   8   7   6   5   4   3   2   1
20 19 18 17 16 15 14 13 12 11

Printed and bound in Great Britain by
CPI Antony Rowe, Chippenham and Eastbourne

For Beth

Contents

List of Tables

List of Figures

Preface

During the 1990s, as part of a global non-governmental forest conservation organisation, I witnessed the rise of sustainable forest management, and the related market-based initiative of timber certification. I was at first hostile to both ideas, as they appeared to confirm my worst fears about sustainable development, namely that it would seek to reduce nature to a tradeable commodity, with no intrinsic, only economic, value. However, at the urging of a colleague, I agreed, somewhat reluctantly, to become a member of the Forest Stewardship Council timber certification programme. Sustainable forest management and timber certification were to have a major impact on my working life over the next few years. In late 1999 when the Australian Federal Government announced that it would be developing its own national certification scheme in collaboration with the forest industry, I was nominated by environmental NGOs to represent their interests. During the same period, a number of timber companies approached me for my advice on how to become certified under the FSC programme. In what seemed a strange twist of fate, given my initial hostility, I had somehow ended up as a major player in the negotiations between economic, environmental and social interests over the introduction of certification domestically.

The advent of certification was to have profound knock-on effects on the ground, and in the following few years it became the main game on the forest scene. In early 2004, when the local state forest management agency announced its intention to log the local valley in which I lived, a collective decision was reached amongst residents that we would try to negotiate a conflict-free settlement. The forest manager, probably mindful, like us, of the protests of previous years, eventually consented to engage in dialogue. After a lot a of discussion, agreement was eventually reached over the final management plan, which confined forestry operations to regrowth forest and protected the remaining old growth in a series of informal reserves.

The success of this encounter led me to ponder what exactly had changed, when previously there had been no other option but protest. And so I decided to find out – a process which has culminated in this book. Clearly, my own direct experience has informed much of my thinking, but this has also been complemented by several years of academic research. Out of this has arisen an analytical framework, which is essentially the product of an organic intellectual. My analysis is firstly an institutional one, and I have sought to synthesise the many and varied theories as to how institutions are, and should be, governed in this era of globalisation. Secondly, I have made use of a range of written and verbal sources to describe the evolution of forest governance over the past two decades. Finally, I have attempted to look at how these developments have been played out within a number of institutional case studies. Of course, every research design has its own flaws, and mine is no exception. I apologise now for any weaknesses or errors, and I encourage all readers to provide me with constructive feedback. In the case studies, it has not been possible to reproduce the level of research required without cluttering the text with, quite literally, hundreds of footnotes. I would therefore refer readers to the bibliography, and suggest they contact me directly if they would like to know more about my sources.

Bellingen,
New South Wales,
A.M.D.G.

Acknowledgements

I would like to thank all those who have assisted me in the preparation of this work. I cannot go any further without honouring my wife and partner, Beth Gibbings, and my son, Tristan Cadman-Gibbings, for their support throughout the course of what has ultimately been a decade-long process. I am deeply grateful to my mentor Dr Fred Gale of the School of Government in the Faculty of Arts at the University of Tasmania, for his insightful comments and assistance at all stages of production. The same goes for Professor Timothy Shaw and all the editorial team at Palgrave Macmillan. I would also like to extend my gratitude to everyone who agreed to be interviewed as part of this study and who gave so generously of their time – including following up conversations and emails. My recognition goes also to Eric Gibbings, for his assistance in explaining how to model three-dimensional concepts two-dimensionally. To all those I have otherwise neglected to acknowledge, I offer my apologies and appreciation. Finally, for the copious amounts of paper used in production of this work, I acknowledge the trees. This book is printed on certified paper, but in deference to the schemes discussed in this volume I have decided not to identify the brand.

List of Abbreviations

ABU	Accreditation Business Unit (FSC)
AFP	Asia Forest Partnership
AFPA	American Forest and Paper Association
AFS	Australian Forestry Standard
ANEC	European Association for the Co-ordination of Consumer Representation in Standardisation
ASI	Accreditation Services International (FSC)
CAG	Chairman's Advisory Group (ISO)
CAR	Corrective Action Request (FSC)
CASCO	Conformity Assessment Committee (ISO)
CBD	Convention on Biological Diversity
CD	Committee Draft (ISO)
CEI	Bois European Confederation of Woodworking Industries
CEPF	Confederation of European Private Forest Owners
CEPI	Confederation of European Paper Industries
CERFLOR	Sistema Brazileiro de Certificação Florestal
CERTFOR	Certificación Forestal en Chile
C&I	Criteria and indicators
CI	Conservation International
CLI	Country-led initiative
CoC	Chain of custody
COFO	Committee on Forestry (FAO)
COP	Conference of Parties (UN)
COPOLCO	Consumer Policy Committee (ISO)
CPF	Collaborative Partnership on Forests (UNFF)
CPPA	Canadian Pulp and Paper Association
CSA	Canadian Standards Association
CSD	United Nations Commission on Sustainable Development
CSR	Corporate Social Responsibility
DCTF	Developing Countries Task Force
DEVCO	Developing Countries Committee (ISO)
DEVPRO	Developing Countries Programme (ISO)

DIS	Draft International Standard (ISO)
EA	European Cooperation for Accreditation
ECOS	European Environmental Citizens' Organisation for Standardisation
ECOSOC	Economic and Social Council (UN)
EEB	European Environmental Bureau
EMAS	Environmental Management and Audit scheme
EMS	Environmental management system (ISO)
ENGO	Environmental non-governmental organisation
EU	European Union
FAO	Food and Agriculture Organisation
FCPF	Forest Carbon Partnership Facility (the World Bank)
FDIS	Final Draft International Standard (ISO)
FERN	Forests and the European Union Resource Network
FFCS	Finnish Forest Certification Scheme
FFIF	Finnish Forest Industries' Federation (Metsäteollisuus r.y.)
FLEG	Forest Law Enforcement and Governance
FNE	France Nature Environnement
FNL	Finnish Nature League (Luonto-Liitto)
FoE	Friends of the Earth
FSC	Forest Stewardship Council
GOF	Global Objectives on Forests (UNFF)
GRI	Global Reporting Initiative
IAF	International Accreditation Forum
ICC	International Chamber of Commerce
IEC	International Electrotechnical Commission
IFF	Intergovernmental Forum on Forests (UN)
IIED	International Institute for Environment and Development
ILO	International Labour Organisation
IMF	International Monetary Fund
INNI	International NGO Network on ISO
IPF	Intergovernmental Panel on Forests (UN)
IPO	Indigenous Peoples' Organisation
IS	International Standard (ISO)
ISA	International Federation of the National Standardising Associations

ISEAL	International Social and Environmental Accreditation and Labelling Alliance
IR	International relations
ISO	International Standardisation Organisation
ITTA	International Tropical Timber Agreement (ITTO)
ITTO	International Tropical Timber Organisation
IUCN	World Conservation Union (formerly the International Union for the Conservation of Nature)
IUFRO	International Union of Forest Research Organisations
LBI	Legally Binding Instrument
LEI	Lembaga Ekolabel Indonesia
MCPFE	Ministerial Council for the Protection of Forests in Europe
MDG	Millennium Development Goals (UN)
MSD	Multi-stakeholder Dialogue (UNFF)
MTCC	Malaysian Timber Certification Council
MTK	Central Union of Agricultural Producers and Forest Owners (Maa- ja metsätaloustuottajain Keskusliitto MTK r.y.)
MYPOW	Multi-year Programme of Work (UNFF)
NEPI	New environmental policy instruments
NFP	National Forest Programme (UNFF)
NGO	Non-governmental organisation
NI	National Initiative (FSC)
NLBI	Non-legally Binding Instrument
NP	New Proposal for a Work Item (ISO)
NPM	New public management
NSMD	Non-state market driven
ODA	Official development assistance (UN)
OLI	Organisation-led initiative
P&C	Principles and criteria
PC&I	Principles, criteria and indicators
PEFC	Programme for the Endorsement of Forest Certification Schemes
PEFCC	PEFC Council
PfA	Proposals for Action (UNFF)
PEOLG	Pan European Operational Level Guidelines
PPP	public–private partnership

PROFOR	Programme on Forests (the World Bank)
PWI	Preliminary Work Item (ISO)
QMS	quality management systems
REDD	Reducing Emissions from Deforestation and Forest Degradation in Developing Countries, post COP 15 referred to as REDD-plus (UN)
REMCO	Council Committee on Reference Materials (ISO)
RLI	Region-led initiative
SAGE	Strategic Advisory Group for the Environment (ISO)
SC	Sub-committee (ISO)
SFI	Sustainable Forestry Initiative
SFM	sustainable forest management
SME	Small and medium enterprise
SMF	Sustainable management of forests
TAG	Technical Advisory Group (ISO)
TBT	Technical Barriers to Trade (WTO)
TC	Technical Committee (ISO)
TFAP	Tropical Forestry Action Plan (FAO)
TFG	Task Force Group (PEFC)
TFRK	Traditional forest-related knowledge (UN/UNFF)
TG	Task Group (ISO)
TMB	Technical Management Board (ISO)
TR	Technical Report (ISO)
UN	United Nations
UNCED	United Nations Conference on Environment and Development
UNCHE	United Nations Commission on the Human Environment
UNCTAD	United Nations Conference on Trade and Development
UNCTC	United Nations Centre for Transnational Corporations
UNDP	United Nations Development Programme
UNEP	United Nations Environment Programme
UNFCCC	United Nations Framework Convention on Climate Change
UNFF	United Nations Forum on Forests
UN-REDD	United Nations Collaborative Programme on Reducing Emissions from Deforestation and Forest Degradation in Developing Countries

WBCSD	World Business Council for Sustainable Development (formerly Business Council for Sustainable Development)
WD	Working Draft (ISO)
WG	Working Group (ISO)
WRM	World Rainforest Movement
WTO	World Trade Organisation
WWF	World Wide Fund for Nature (World Wildlife Fund)

1
Introduction

Origins of global governance

The institutions that emerged in the immediate post-Second World War period, most notably those associated with the United Nations (UN) Bretton Woods Conference, represent an important step in international economic, social and political relations. Previously, conflicts between nations were resolved principally by means of the Westphalian model of international diplomacy, whereby state-centred sovereignty remained unchallenged. Today, this sovereignty still remains intact, but it is increasingly the subject of a system in which governments are rendered accountable to each other – albeit imperfectly – through such bodies as the UN. A whole series of covenants, treaties and declarations has also arisen between contracting states in the wake of the post-War settlement, including the 1948 Universal Declaration of Human Rights. Despite their shortcomings, these agreements represent an attempt to redefine the national interest and protect society's minority groups from the excesses of the state – and from each other. The result is a far more global understanding of international relations (IR), institutionally expressed through a range of such bodies as the Food and Agriculture Organisation (FAO) and the International Monetary Fund (IMF). These developments, combined with the integration of global financial activities in the post-Cold War period and the ongoing growth of information and communication technologies, have all contributed to the description of the contemporary era as one driven by the processes of globalisation.

At the same time, and historically commensurate with the rise of globalisation, it is also possible to discern an increasing 'environmentalization' of global institutions. Two important contributions were the establishment of the International Union for the Conservation of Nature and Natural Resources (IUCN, now the World Conservation Union) in 1948 and the creation of the United Nations Commission on the Human Environment (UNCHE) 20 years later. The United Nations Environment Programme (UNEP) and the Stockholm Declaration of 1972, both of which arose out of UNCHE, placed the imperative for environmental action on the global level, and set the normative context for future discussions about the environment. Global action reached a high point with the United Nations Conference on Environment and Development (UNCED), held in Rio de Janeiro in 1992.

The substantive outcome of Rio, Agenda 21, formally recognised the participation of non-state interests, particularly non-governmental organisations (NGOs) in the framework of international environmental policy and environmental decision-making at all levels.[1] Participation is an essential component of citizen power, without which democracy is an empty and frustrating process.[2] Despite this degree of acknowledgement, governments still see themselves as the chief political negotiators at the global level, and democratic participation – if it occurs at all – continues to take place within the nation-state. Global decision-making remains largely in state hands, undertaken between governments meeting in forums that exist beyond the electoral mandate of national citizens. But these historical arrangements are under pressure as a result of the globalisation and environmentalisation of world politics. The supra-national nature of the UN, and other regional projects such as the European Union (EU), are challenging the very idea of national sovereignty, even if that sovereignty remains formally intact. Furthermore, contemporary issues, climate change being the chief exemplar, are not always contained within territorial borders. These developments have impacted on the relationship between the individual and government, and national citizens are questioning the political supremacy of the nation-state. This makes for a complex debate regarding roles and responsibilities in the postmodern era, which is characterised by its complexity, uncertainty and risk. The expectation for increased citizen participation that these developments have brought about pose major problems for contemporary democracy, not only in terms of how to structure institutional responses in ways that effectively deal with global problems, but also in terms of how newly enfranchised actors should be included in decision-making processes. Territorial systems, in which political participation is limited almost exclusively to electing representatives to make decisions, are no longer the only sites of democracy.

As a result, there has been an ongoing evolution away from traditional processes of govern*ment* towards the more abstract concept of govern*ance*, based on based on the 'dynamic interplay between civil society, business and public sector'.[3] But this transition from government to governance is neither straightforward nor uncontested. Alternative institutional venues have not supplanted the sovereignty of the nation-state. Some complement existing intergovernmental processes, others may compete with them. This competition is made all the more significant in view of claims that traditional multilateral institutions are weak, and have proven to be of only limited value in solving global problems.[4]

Understanding contemporary governance

It is important to understand the evolution away from government and towards governance that has arisen in the wake of globalisation. In explaining contemporary developments, conventional disciplines are no longer entirely applicable in their classical form. This is particularly the case with IR. The previously orthodox viewpoint that geopolitical cooperation occurs almost exclusively within the sphere of intergovernmental regimes comprised of intergovernmental agreements that are pursued in the context of state-based authority is now considered as being largely out of touch.[5] The regime concept, particularly influential in recent theory, is beginning to be replaced by the more relevant idea of multi-level governance. Governance itself is also becoming increasingly understood in terms of its expression not only on the national and international levels, but at all spatial scales. Contemporary environmental governance articulates this trend particularly strongly, and is exemplified by the interactions that occur between decentralised networks made up of multiple actors functioning at all levels, from the global to the local, and vice versa.

A body of theory in the field of comparative politics has also arisen since the 1990s, which argues for a broader understanding of state and non-state relations than those explained by traditional top-down, command-control models of regime-based state authority. Modern governance is portrayed as essentially social–political in nature, and understood as ongoing processes of interaction between social groups and forces within public and private institutions.[6] Interaction is key, and is identified as a series of 'co'-arrangements between state and non-state actors, more oriented towards collaborative approaches to problem-solving based on the formulation of criteria, or the setting of standards.[7]

In such models the transmission of information and knowledge and its valuation by those involved plays a central role; and deliberation, as opposed to directive, has become an alternative mechanism of dealing with the complexity and ambiguity of political and social problems.[8] Deliberation has come to be identified as a specific method of political interaction, markedly different from the traditional practice of democracy, as it is seen within nation-states, and especially when this practice is extended into the international arena. Deliberation occurs when problems are discussed with a view to developing solutions through cooperation and joint agreement and in which rational discourse contributes to problem-solving.[9] This is to be contrasted with the traditional democratic practice of aggregation, or aggregative democracy, when divergent interests are grouped together and the one with the greatest level of influence (often numerical) prevails. This is the most common system on the national level. The designation of democracy as deliberative or aggregative is a critical theoretical method of interpreting the particular expression of a given system, and in the final analysis, is characterised as consisting of cooperative versus competitive political interaction.

These new systems now sit alongside traditional, more legalistic, mechanisms and have been interpreted as representing new processes of governing.[10] There is also a growing acceptance that governance theory and analysis is grounded on the assumption that structure and process is fundamental to understanding the quality of interactions between participants in contemporary global institutions.[11] This emerges in the material of the early 1990s, and in the light of research experience re-emerges a decade later in terms of 'governance as structure', understood as the models utilised by various institutions, and 'governance as process', referring to the idea of steering or coordinating, and comprising the 'co'-arrangements referred to above.[12] Together, these two conceptions of governance have been identified as the key determinants of 'governability', understood as the overall capacity of a system to govern itself.[13] In the light of such conceptual developments, and in order to make governance more effective, there have been calls for researchers to think about institutional design more creatively.[14] Following on from previous work, this book establishes an institutional relationship between *participation as structure* and *deliberation as process*. In this conception, participation and deliberation have a functional significance beyond their particular expression in a given institution; it is not the institution per se, but rather how participation and deliberation occurs within it that determines the effectiveness, or quality, of its governance.

But even if quality of governance is conceived of in these terms, it is nevertheless still necessary to determine what makes an institution legitimate, since there is disagreement between governance theorists as to whence legitimacy is derived. Two theories currently dominate. Legitimacy can be 'input oriented': that is, derived from the consent of those being asked to agree to the rules, and concerning such procedural issues as the democratic arrangements underpinning a given system. Legitimacy can also be 'output oriented': derived from the efficiency of rules, or criteria for 'good' governance, and demonstrated by substantive outcomes.[15] Legitimacy can therefore be determined both according to the principles of democracy on the one hand and efficiency and effectiveness on the other. Input legitimacy concerns itself with the structures and processes of governance, whilst output legitimacy is more interested in results. Additionally, recognising the social–political nature of contemporary governance emphasised in the literature, it is also necessary to further conceive of legitimacy in sociological terms.[16] In this broader context, legitimacy of governance should be understood in terms of the social–political interactions within the structures and processes of an institution and the outcomes they generate; the more balanced these elements, the more governable the system.[17] It is the quality of these interactions that ultimately determines legitimacy. This interrelationship can be expressed figuratively (see Figure 1.1 below).

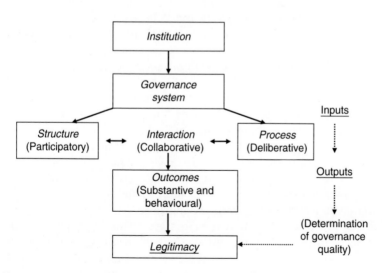

Figure 1.1 Conceptual model of contemporary global governance.

Institutional expressions of governance

Given the diffusion of power away from the old paradigm of top-down, command and control state-led approaches, there has been a proliferation of new forms of governance and associated policy instruments. Starting at the shift in global politics away from intergovernmental, state-based relations, it is now possible to discern a growing trend towards alternative forms of governance. This is a development most analysts no longer deny. Such forms are characterised by their network-like relations between private and public actors and include such arrangements as business-driven self-regulation, and public–private or civic–private partnerships.[18] They are closely related to economic globalisation and the reduction in state power, as well as the weakness of global governance at intergovernmental levels. Ironically, the move to private sector governance has been given a boost as a result of the international institutions that have either arisen within the UN, such as the UNEP, or via UN-sponsored initiatives, such as the Global Compact (2000). But UNEP and other initiatives notwithstanding, the UN system has been criticised as being effective neither in the assessment, review and monitoring of the measures it prescribes, nor in ensuring compliance.[19] As a result, although corporations still lobby governments through traditional processes based around multilateral agreements, they are also dealing with each other and other non-state actors as well as the state. Such efforts amongst private actors, civil society and the state have given rise to private and civic governance arrangements that resemble the public governing functions of states and intergovernmental institutions. Relations are of a permanent and institutionalised nature and should be distinguished from simple cooperation, which is ad hoc and short-lived.[20]

Such private and civic initiatives are categorised in a number of ways. Self-regulation is defined as 'mechanisms to reach collective decisions about transnational problems with or without government participation'.[21] Three discrete self-regulatory models have been described, centred upon civil society-based monitoring and advocacy initiatives, industry-driven standard setting, and traditional government-led regulation. Self-regulation should not simply be interpreted as industry opportunism seeking to undermine regulatory systems to the lowest common denominator. Many companies are in fact developing international standards that exceed the requirements of national legislation. Instead, self-regulation should be interpreted as further evidence of the changing nature of global governance.[22]

Although self-regulation is seen as having the potential to encourage significant improvements, a range of other traditional and newer political approaches is in use. Indeed, there is an interplay between governance, government and policy instruments, which reflect the nature of the changing roles of state and societal actors, in a period of what has been referred to as governance transition. In the literature covering environmental policy instruments particularly, a trend away from the traditional regulatory approach dating back to the late 1960s towards new environmental policy instruments (NEPIs) has been noted. Various sub-types have been identified, but there are three of particular significance: voluntary agreements, environmental management systems (EMS) and market-based instruments (including eco-labels). Older policy instruments continue to compete with these NEPIs, which are themselves competing with each other. Hybrid forms of regulation, with strong features of governance such as self-monitoring and societal organisation, are also appearing. Governance and government therefore represent a continuum of different governing types on two heuristic 'poles'. On balance, though, jurisdictions have nevertheless shifted from a position of government to governance with regard to their use of newer instruments.[23]

The role of non-state actors is central to these newer instruments. New sub-political arrangements have emerged that are institutionalising environmental, non-state interests in the economic domain, especially when governments fail to act, or when intergovernmental agreements and institutions prove inadequate. Market-based mechanisms are making an increasingly significant contribution to global reform in this regard.[24] An initial wave, brought about by public pressure, consisted of largely unilateral company codes, designed to demonstrate good conduct, but not generally open to public scrutiny. These instruments were intended to address the concerns of consumers in industrialised countries concerning issues such as child labour, but ignored more entrenched problems, such as freedom of association. This was followed by a further development with greater levels of scrutiny, based around reporting, and designed to demonstrate a company's environmental and social responsibility to its shareholders. The third and most ambitious was the creation of sector-wide arrangements, involving several businesses and/or associations, and including civil society participants. All three used certification of one form or another to verify compliance. While the resultant institutions that have arisen do not and cannot replace nation-states, they have introduced new elements and dynamics into the processes of global governance.[25]

Classifying governance

A further and more complex result of the evolution of contemporary governance is its diversity, and the ways in which it is categorised in the literature. Theorists dispute the number and types of contemporary governance. Most authorities are able to see a difference from earlier conceptions, generally described, as noted earlier, as being top-down in nature; contemporary governance by contrast is portrayed as being less command-and-control oriented, but there are disputes over the phenomenon of governance, which can make comparisons between theories difficult. These differences depend both on theoretical perspective and date, and can generally be summarised according to scholarly discipline – but by no means exhaustively – as in Table 1.1 below.

But rather than presenting global governance as existing only within rigid definitional sets it would be better to conceive it as a dynamic interplay between the factors influencing institutional expression. Three major factors in particular stand out amongst the many issues affecting the practice of global governance, and the institutions in which it is expressed. The first of these is that a shift in locality away from the nation-state to multiple sites implies that the nature of contemporary *authority*, or sovereignty, is changing. Authority may be located within national governments, but, given the nature of global social and economic transactions, it may also be vested in non-state agents, from corporations (or alternatively formulated, the private sector) to NGOs (or civil society), and multilateral organisations such as the World Trade Organisation (WTO), IMF and so forth; even traditional notions of public and private are no longer clear-cut. Here, the old state-centric exercise of authority, and the new power of non-state organisations, exist on two ends of a continuum. The second and related theme, given the erosion of the role of the nation-state as the sole sphere of authority, regardless of its continuing existence and contribution to global politics, is the discussion across the literature about the practice of *democracy* in a globalised world. In the old world, democracy is characterised by territorially located, political parties; in the new, non-territorial sectors, consisting of groups such as civil society and business, engage one with another in a much more cooperative set of arrangements, where collaboration is central to rule-making. Here, deliberation is to be contrasted with the other democratic end of the continuum, where interests are aggregated and compete with each other. Thirdly, within governance theory itself, the discussion is also about the move away from traditional, towards new forms of governance.[26] This is complicated by the fact that it is easier to distinguish between 'old' than the many and varied 'new'

Table 1.1 Four typologies of governance 1997–2008

Public Policy	Form	International Relations	Form	'Analytic' (Arts)	Form	Comparative Politics	Form
Centralised	Government has control	*Top-down*	Traditional governmental & intergovernmental relations (including business)	*Old*	State-steered (top-down, command-control)	*Private transnational*	Financial market schemes
Minimal state	Less government, more privatisation	*Bottom-up*	Informal civil society initiatives	*New*	New modes (self-regulation, etc. – public-private)		NGO programmes
Corporate	Directed and controlled by companies	*Market*	Multiple players using formal & informal market-based mechanisms	*All*	New and old mechanisms for procuring public goods (public, private & mixed)		Business/ NGO partnership
New public management (NPM)	Private sector practices in the public sector	*Network*	Formal state, civil society, business alliances	*Normative*	Programmes to renew management (good governance, NPM & corporate governance – public & private)		Multi-stakeholder processes

Table 1.1 (Continued)

Public Policy	Form	International Relations	Form	'Analytic' (Arts)	Form	Comparative Politics	Form
'Good'	Practices of NPM and liberal democratic values	Side-by-side	Informal cooperative arrangements between state & non-state				International Union based projects (codes of conduct; business partnerships)
Socio-cybernetic	Social-political interaction	Mobius-web	Intricate, overlapping mixed arrangements ('end-state' of contemporary governance)				
Self-organising Networks	Interdependent actors/agencies delivering services						

Sources: Rhodes (1997), Rosenau (2003), Arts (2006), Dingwerth (2008).

governance types.[27] In practice, divergent forms of governance also appear alongside each other in the global policy arena. This is demonstrated, for example, by the existence of intergovernmental multilateral environmental agreements, exemplified by the United Nations Forum on Forests (UNFF), alongside non-state market-driven instruments, such as the Forest Stewardship Council (FSC). Consequently, *innovation* provides a third benchmark against which contemporary governance is to be understood.

These three factors are interpreted here as influencing the type of governance expressed in a given institution. Each institution will also sit in different places along the continuum in relation to each of these factors. This interplay can be expressed conceptually by way of a three-dimensional attribute space. By allocating some form of simple rating system (for example, low, medium and high) it is also possible to determine the extent to which these themes are expressed in existing institutions, and to plot those institutions in three dimensions (see Figure 1.2 below).

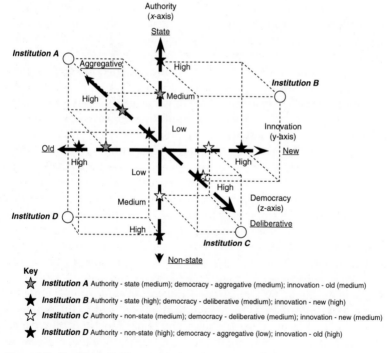

Key

☆ **Institution A** Authority - state (medium); democracy - aggregative (medium); innovation - old (medium)

★ **Institution B** Authority - state (high); democracy - deliberative (medium); innovation - new (high)

☆ **Institution C** Authority - non-state (medium); democracy - deliberative (medium); innovation - new (medium)

★ **Institution D** Authority - non-state (high); democracy - aggregative (low); innovation - old (high)

Figure 1.2 Typological framework for the classification of four hypothetical governance institutions.

Sources: Adapted from Koenig-Archibugi (2006), Reuben (2003).

Major elements of governance

The term 'governance arrangement' has come to be used to refer to a range of specific mechanisms influencing the nature of the interaction between the participants involved collective problem-solving.[28] A wide range of scholars has commented at length on the various arrangements underpinning governance. These institutional elements, identified across the fields of governance theory, have a bearing on governance quality and legitimacy, and are worth exploring at this point.

Interest representation

Interest representation, closely identified with the inclusion or inclusiveness of stakeholders, is recognised as a fundamental problem in contemporary global governance and a source of considerable institutional variation. Inclusiveness has been broken down into two elements, access and weight. Access concerns those who may be bounded or affected by a given policy and whether they actively participate in developing regulatory responses, whilst weight denotes the extent to which influence is shared equally amongst participants. Inclusiveness therefore sits along a power continuum and is measured by the extent to which participants are involved in decision-making processes and whether their input is taken into account.[29] Democratic theorists commenting on global governance also link equality to inclusiveness – and by extension, exclusiveness to inequality – arguing that legitimacy is normatively expressed by participation in decision-making. Processes at the global level in which nation-states play the predominant role are inherently exclusive and in order to increase levels of representation require a wider range of previously marginalised groups and perspectives.[30] There is an inherent, and problematic, interplay between interest representation, inclusion and equality in these kinds of non-spatial and non-territorial venue. Simply using previous nation-state oriented democratic standards can erode equality even if there are increased opportunities for involvement.[31] Although many other theorists accept that participation within contemporary governance is basically representative and group-oriented in nature, they likewise recognise that the complexities brought about by globalisation call for a reinvention of interest representation. Attention has consequently been paid to the problems associated with greater participation and wider inclusion inherent in the governance concept, notably the prevalence of hidden places of influence and power.[32] This problem has been referred to elsewhere as 'fugitive' power and can result in a number of governance legitimacy issues; this led the European Commission, for example, to identify inclusiveness as one of its principles of 'good' governance in a White Paper in 2001.[33]

Effective interest representation in global governance also requires significant resources. These are generally only available to well-endowed organisations, residing in the more privileged parts of the world. Lack of resources can be offset when associations collaborate, however, and in this regard at least, networks have been identified as having the potential to play a beneficial role. If the problems surrounding how networks are themselves resourced can be overcome, such structures have the potential to impact positively on global governance.[34] Scholars further point to the need for economic-technical capacity (money and expertise) as well as institutional support as a structural framework condition for organisations seeking to develop effective policy within decision-making processes that include both public and private actors.[35]

Accountability and transparency

Accountability has become a central aspect of the quality of governance debate, since the rise of new actors and new institutions has necessitated a reconfiguration of existing democratic mechanisms for holding participants to account.[36] It is seen as being both an internal and external requirement of good governance.[37] There is a need for a better meshing together of internal and external accountability measures, posing a number of unresolved normative questions.[38] Some scholars, seeing that defining what constitutes a robust accountability system is a major problem facing advocates of new governance, have problems envisaging any serious contenders to the state as a source of democratic accountability.[39] Consequently, there have been calls for properly adapted principles and standards of accountability to satisfy normative democratic criteria as world politics generally lack universally accepted values and institutions. Greater freedom of information might compensate for the remoteness of global processes from democratic accountability.[40] In this context, the external accountability of decision makers is to the public at large, and is linked to what appears as a related attribute, transparency, expressed in terms of public access to information and decision-making procedures.[41] Transparency plays a role both in the participation of interests from the inception of a governance system or policy process (*ex ante*) and in the public scrutiny of decision-making (*ex post*).[42] It is effectively a precondition for effective accountability, since it is impossible to hold an institution to account if its regulatory operations are not open to public view.[43] Formal structures and clearly defined rules are required for each level, otherwise transparency can be lost, and policymaking predetermined.[44] In short, how the responsibility of participating actors should be addressed in non-state, non-elected structures continues to be the subject of some debate.

Decision-making

There is general recognition that as governance continues to develop at a global level, procedural arrangements are likely to be based more and more upon commonly agreed rules and practices.[45] Governance itself is increasingly multi-level, undermining traditional concepts of community and representation, generating a form of decision-making, which is occurring in forums that in addition to their democratic expression are also, as indicated above, more deliberative in procedural style.[46] Current intergovernmental systems are seen as lacking the necessary processes to cope with greater degrees of non-state participation. This has led to the conclusion that without existing institutional arrangements being changed in favour of more productive interaction, built around consensus, global environmental negotiations will continue to produce inadequate results.[47] EU regime scholars have examined whether decision-making operating on a consensual or majority basis is more effective.[48] Anecdotal case studies of environmental processes, echoing those of their EU counterparts in the US, indicate that agreement is often reached by consensus (understood as total agreement) during the working stages of negotiation, reverting to a majority vote at the end.[49] However, commentators critical of current policymaking have noted an almost obsessive trend to consensus within new modes of regulatory governance. Criticisms include the definitional inconsistencies of consensus, which can be defined as both constituting unanimity, or as a decision everybody can live with.[50]

Policy-making, implementation and enforcement inevitably involve conflict amongst interested parties requiring dispute-resolution mechanisms. The inability to resolve conflicts has been identified as a key indicator of governance failure. Interestingly, it is in the arena of environmental governance that some of the most extensive use of these processes is made. In cases of environmental dispute resolution, it has been noted that the consensus developed through mediation can bring separate interests closer together.[51] Such an approach would again consist of 'discursive procedures' for dispute settlement through the solving of problems cooperatively.[52] Conflict should therefore not be seen as a negative aspect of governance. So long as it does not encompass irreconcilable issues, such as matters of religion or ideology, for example, conflict can be managed, even if it is ongoing. Indeed, the ongoing nature of 'muddling through' a particular conflict may even set the stage for the next round of engagement and negotiation.[53]

Implementation

In order to determine whether a given policy objective has been implemented effectively it is necessary to trace the final effects of a given policy and its related programmes.[54] EU scholars stress the relationship between implementation and compliance. If implementation is the action of putting commitments into practice, compliance relates to the consistency between a given rule and an affected party's behaviour. In this context, effectiveness is presented as a measure of the extent to which a policy has been successful in solving the problem it was created to address. Compliance is consequently seen as a proxy for effectiveness.[55] Successful implementation therefore relates to both behavioural and problem-solving effectiveness.[56] The importance of problem-solving as an aspect of institutional governance has led one scholar to note: 'An environmental regime is successful when it solves the problem that led to its creation.'[57] Notably, it is argued, an institutional approach to problem-solving that incorporates a degree of flexibility results in governance systems that are more resilient in the face of external change and may even benefit from it. Non-resilient systems, on the other hand, are vulnerable to change.[58] Here, the implication is that flexibility and adaptability contribute to the durability of an institution. However, it is important to bear in mind that factors external to the governance system also impact on the success of implementation. Governance systems stand little chance of improving situations where legal requirements and enforcement capacities are weak, and where social, economic and political contexts beyond the institution itself impede successful implementation.[59]

Evaluating governance quality

Although a wide range of scholars has commented at length on the various attributes of 'good' governance, as the discussions above demonstrate, they have generally not examined the nature of the relationship between those arrangements as a whole, as they often focus on a single criterion (the most frequently discussed being accountability). If they do comment on several criteria, these discussions are more often related to the authors' areas of expertise, or particular institutional focus, than on the significance of these criteria as parts of a holistic mechanism contributing to quality of governance. In addition, the terminology used for describing these governance arrangements is inconsistent and criteria and indicators (C&I) are used interchangeably. If there are inconsistencies in the academic literature, it should come as no surprise to see this reflected in practice. The lack of a consistent and coherent approach to measuring and verifying governance quality has meant that institutions face considerable difficulties in reaching agreement over the necessary elements for assessing the legitimacy of competing systems.

For the purposes of understanding the nature of the relationship between the various governance arrangements discussed above, two simple methods for evaluating governance quality have been developed for the case studies, which follow. The first is a framework of principles, criteria and indicators (PC&I) relating to governance quality. A *principle* is a fundamental rule, which serves as a basis for determining the function of a complete system in respect to explicit elements. A principle can also express a certain perspective, or value, regarding a specific aspect of the system as it interacts, in this context, with the overall governance system.[60] *Criteria* function at the next level down, and can be described as categories of conditions or processes, which contribute to the overall principle.[61] They are intended to facilitate the assessment of principles that would otherwise be ideational and non-measurable. Criteria are not usually capable of being measured directly either, but are formulated to provide a determination on the degree of compliance. They are consequently linked to *indicators*, which are hierarchically lower, and which represent quantitative or qualitative parameters, and do describe conditions indicative of the state of the governance system as they relate to the relevant criterion. Together, PC&I may be used as the basis for *standards*, which serve as a reference for monitoring, reporting and evaluation.[62] Standards are the substantive outcomes of the governance system and determine how the system is formulated and applied, thereby delivering effectiveness and legitimacy.[63] The relationship between PC&I, and the ways in which the various governance arrangements discussed above can be

Table 1.2 Hierarchical framework of principles, criteria and indicators of governance quality

Principle	Criterion	Indicator
'Meaningful participation'	Interest representation	Inclusiveness Equality Resources
	Organisational responsibility	Accountability Transparency
'Productive deliberation'	Decision-making	Democracy Agreement Dispute settlement
	Implementation	Behavioural change Problem-solving Durability

formulated for assessing institutional performance, are laid out in Table 1.2 above.

Here the rule (perspective, value) concerning participation is that it should be *meaningful*. This term is frequently associated with participation in much of the literature, and serves here as a normative, qualitative descriptor.[64] The second principle, referring to the deliberative, procedural aspects of governance, has been ascribed the term *productive* as its descriptor.[65] This refers to the quality of deliberations, as they occur within the system, as well as the quality of the outcomes, or products, of those deliberations. Meaningful participation is assessed through two criteria, *interest representation* and *organisational responsibility*. In the discussion of the governance literature presented above interest representation is seen as a key component of 'good' governance, and has been linked to three elements of governance, which function on the indicator level: *inclusiveness*, demonstrating who has access to a governance system; *equality*, indicating the balance of power, or weight, of participants; and *resources*, referring to the economic, technical or institutional capacity of a participant to represent their interests within the system. The second criterion, *organisational responsibility*, relates to whether the range of organisations involved in a governance system can be considered as acting responsibly. This includes the institution itself, related entities (such as accreditation and certification bodies) and its constituents (members and non-members). Responsibility comprises two indicators, *accountability* and *transparency,* which are usually treated together in the literature, and refer to the extent to which the behaviour of participating organisations can be called to account both inside the institution and externally by the public at large, and the degree to which their behaviour is visible and open to scrutiny by other actors within and beyond the institution.

The procedural principle of productive deliberation is demonstrated through two criteria, *decision-making* and *implementation*. Decision-making is an essential part of the deliberative process, and is linked in this study to three indicators: *democracy*, not referring to a specific mode of democracy, but rather the extent to which a system can be deemed to be functioning democratically; *agreement,* referring to the method in which decisions are reached, such as voting, or consensus; and *dispute settlement*, indicating the system's capacity to manage conflict when there is no agreement, or there are challenges to decisions made. Implementation refers to the process of putting commitments into practice.[66] However, the fact that a system has created policies or standards does not automatically demonstrate productiveness. Three indicators are therefore associated with implementation in this study: *behaviour change*, used to determine whether the implementation of agreements, or substantive outcomes results in changed behaviour regarding the problem that the system was created to address; *problem-solving*, referring to the extent to which the system has solved the problem it was created to address; and *durability*, capturing the related elements of adaptability, flexibility and resilience.

Scope of this book

The central argument of this book is that *the more deliberative and participatory the governance system the higher its quality and the more legitimate the institution*. Institutional use of aggregative models of democracy invariably pits social, economic or environmental needs against each other, as divergent interests must compete for supremacy one over the other. Grounded as they are within the territorial confines of the nation-state, such institutional approaches are no longer adequate in an era of globalisation, in which the need for collaboration to tackle global problems has become a pressing necessity.

Deforestation, the central environmental problem under scrutiny in this book, serves as a case in point. Although almost everyone at this historical juncture is prepared to accept there is a problem of over-extraction, national governments continue to face considerable political difficulties in determining how to best manage forests when they are used by many individuals in common, and there is conflict over what action to take. Institutions seeking to combat deforestation and regulate forest management also exist within a global market system that continues to ignore the existence of environmental limits; it compels ever-increasing production in a world that is limited. As a consequence, and despite their efforts, global deforestation continues to escalate.

A key objective of a number of the institutional responses to combat deforestation is to promote *sustainable forest management* (SFM), which is the process of managing a forest to ensure ongoing production whilst maintaining its environmental and social functions.[67] This objective has been pursued via intergovernmental initiatives and private sector programmes, such as forest certification, a major focus of this study. *Certification* in this context should be understood as 'a process, which results in a written quality statement (a certificate) attesting to the origin of raw wood material and its status and/or qualifications following validation by an independent third party'.[68]

Deforestation has been attributed to a wide range of causes including international development and debt policies, population growth, poverty, drug cultivation, wars and the role of the military, land tenure inequities and natural causes. Transnational corporate financial investment associated with the exportation of timber from producer countries to consumer countries has also been identified as playing a significant role.[69] As the subsequent case studies demonstrate, neither SFM nor certification have been entirely successful policy mechanisms for tacking deforestation. Deforestation is often the result of non-forestry activities, such as clearing land for cash-crop agricultural commodities like palm oil. In this case, no matter how exemplary the management of a given forest may be, if the market dictates a higher price for palm oil than timber, the incentive for conversion can be overwhelming for even the best manager. It is therefore somewhat simplistic to interpret SFM as the 'solution' to the 'problem' of deforestation.[70] Rather, the significance of SFM lies more in its contribution to the discourse of sustainable development – and how this discourse is implemented institutionally – than in being a universal panacea for deforestation. In the case of market-based approaches, the point has been made that certification, for example, has a limited impact on the problems of forest degradation and deforestation. From this standpoint the contribution of each of the case studies should be seen as partial, and relating more to the extent to which what they have to offer does in fact complement the range of other initiatives aimed at improving forest management practices.[71]

Reflecting the complexity of contemporary governance generally, forest governance is expressed in a number of different models, both state and non-state. Four global institutions of forest governance have been selected to explore the book's central argument. Each of the four schemes has been selected on the basis of their different arrangements for interaction between social, economic and environmental interests on the global, national and local levels (both public and private, or state and non-state), the model of governance they employ, and democratic mode. The case studies are the FSC, the Programme for the Endorsement of Forest Certification Schemes (PEFC), the International Standardisation Organisation (ISO) 14000 Series (EMS), and the UNFF.

The FSC is the first institution investigated, and, as a system of forest certification, it is particularly interesting as it represents an early 'ideal type' of non-state market-driven governance on account of its strongly non-governmental (as opposed to governmental, or intergovernmental) orientation regarding rule-making authority, standard setting and compliance verification.[72] This is reinforced by the fact that it has been created largely by civil society, with some contribution from business.[73] PEFC provides an interesting counter-example to FSC, since like FSC it was largely initiated originally by a single interest grouping, but in its case forest owners, as opposed to NGOs, and is an essentially 'business' model of governance. How it expresses the various institutional attributes identified in this study as contributing to quality and legitimacy provides some interesting insights into the quality of this type of forest governance versus the more 'civic' model of FSC. PEFC provides an umbrella under which national certification initiatives challenge FSC's 'original' model of market-driven governance. The differences between these two models consequently provide further insights into the broader understanding of the strengths, and weaknesses, of certification as a regulatory tool for forest management.

The ISO 14000 Series does not deal specifically with forest management, although its standards can be applied to both forestry and forest products. There were plans for the development of a forest-management specific standard, but these were later abandoned, an interesting story in its own right. Because these standards were developed within a broader institutional context, it has at times been necessary to look both at ISO itself, and the technical committee (TC 207) under which the 14000 Series was developed. As a result, the assessment of the governance quality of the 14000 Series cannot be viewed in isolation from ISO as a whole, and the evaluation of the Series consequently acknowledges this institutional context. It has been included on account of its value as another global environmental governance system with an emphasis on environmental management, and an associated certification programme.

UNFF provides an example of an existing – intergovernmental – forest-related model. It has been chosen for the express purpose of rounding out the comparative analysis, contained in Chapter Seven, by enabling the reader to compare non-state, market-based systems with a more conventional, state-based approach to governance. With no formal standards development process, and no related certification scheme, it differs considerably from the other case studies in this regard. As an intergovernmental body, most of its activities are concerned with UN member countries and their national forestry-related regulations and programmes. It concentrates less on the role of the market as an environmental problem-solving mechanism, and places more emphasis on forest policy instead. Consequently, its approach to participation and deliberation is viewed more in terms of the sovereignty of the nation-state, than the approaches of the other systems examined. This orientation produces a different expression of contemporary global governance, but one that is equally suited to the analytical methods used in the preceding three studies.

Of the four institutions the largely non-governmental initiative, FSC, achieves the highest rating. The point to be emphasised is that it is not any inherent superiority of FSC that merits this evaluation, but the quality of the structural and procedural interactions within its governance system. It is the degree to which participation as structure and deliberation as process is institutionally embedded that ultimately determines legitimacy.

Method

Each of the institutions has been subjected to a review of primary documentation from the institution itself, secondary materials from academic scholars, and 'grey' literature generated by NGO commentators and consultants. Interviews were also conducted with representatives from the main interests – economic, environmental, social and governmental – that participate in forest governance, as well as with staff from each of the institutions. Given the size of the sample (37) it was not possible to divide interviewees along strictly sectoral lines without revealing the identity of individuals, so participants have placed into three groups: NGOs (social and environmental); Business (private consultants, company employees and NGOs with a specific business orientation); and 'Other' (institutional and governmental representatives, and other agencies). A full list of interview participants and their designation is included in the Appendix. These sources have been used to evaluate the case studies that follow. The intention behind using three discrete sources of information for each case study (primary and secondary sources and interviews) is to provide for a 'triangulated' critical analysis, not reliant on a single source.[74] It should be stressed, however, that the 'score' is a qualitative one; readers can decide for themselves whether to accept this evaluation or not.

The data thus collected have been used as the basis for qualitative analysis of institutional performance, generating firstly, an institutional classification following the typological framework of Figure 1, and secondly, the hierarchical set of PC&I in Table 1.2 above. Each of the institutions is also rated using a Likert scale, ranked from very low to very high. Performance is evaluated at the indicator level, which for the purposes of comparative analysis is also recorded in numerical terms (from one to five points) with a 'pass' threshold of 3 (or 'medium' rating). Following the hierarchical assessment framework of PC&I, the cumulative values of the relevant individual indicators demonstrate the degree of fulfilment at the criterion level; these criteria in turn form the cumulative basis for determining compliance at the principle level; at both the criterion and principle levels a value of 50 per cent is used to determine the pass/fail target. Results are to the nearest percentile. See Table 1.3 below.

Contents outline

Having explored the origins, development, expression and evaluation of contemporary governance in this Introduction, Chapter Two sets the historical and institutional scene for the case studies, which follow. The chapter traces the historical development of international cooperation over forest management prior to and after the 'Earth' Summit. It argues that traditional multilateral processes were ultimately incapable of solving the forest problem, leaving the way open for alternative market-based mechanisms, such as the FSC and other more nationally focussed certification initiatives. The next four chapters systematically investigate the governance arrangements of each of the case studies, commencing with FSC, and following with ISO, PEFC and UNFF. After a

Table 1.3 Evaluative matrix of institutional performance

Principle	1. Meaningful Participation					
Criterion	1. Interest representation Highest possible score: 15 Lowest possible score: 3			2. Organisational responsibility Highest possible score:10 Lowest possible score: 2		Sub-total (out of 25):
Indicator	Inclusiveness	Equality	Resources	Accountability	Transparency	
Very high	5	5	5	5	5	
High	4	4	4	4	4	
Medium	3	3	3	3	3	
Low	2	2	2	2	2	
Very low	1	1	1	1	1	

Principle	2. Productive deliberation					
Criterion	3. Decision-making Highest possible score: 15 Lowest possible score: 3			4. Implementation Highest possible score: 15 Lowest possible score: 3		Sub-total (out of 30):
Indicator	Democracy	Agreement	Dispute settlement	Behaviour change	Problem solving	Durability
Very high	5	5	5	5	5	5
High	4	4	4	4	4	4
Medium	3	3	3	3	3	3
Low	2	2	2	2	2	2
Very low	1	1	1	1	1	1
Total (out of 55)						Final Score:

history of the emergence, evolution and significant controversies affecting each particular case study, a second section presents an overview of its governance system and type, followed by an evaluation of the institution's performance. Each chapter finishes with some case-study specific conclusions. Chapter Seven provides a systematic comparison of the governance arrangements of all four case studies, highlighting similarities and differences and discussing the relationship between institutional type and performance. Chapter Eight, the conclusion, summarises the case lessons, revisits the analytical framework, and ends with a commentary on some of the salient developments in governance theory since the completion of research.

2
Governance and Forest Management

Forest governance provides one of the best spaces available to study the emergence of the new modes of governance that have arisen in response to globalisation.[1] It is in the forestry arena that environmental governance, understood as 'the coordination of interdependent social relations in the mitigation of environmental disruptions' most clearly reflects the involvement of civil society and private industry, as well as the state, in the development of regulatory regimes.[2] Some of the most extensive and innovative experiments in 'new' governance – of which forest management certification is one of the most interesting – are to be found in the sector. The policy responses generated within this arena in reaction to the larger political and economic developments associated with globalisation are also interesting from a theoretical perspective, and yield some interesting insights into the evolution of global governance more broadly. The comment has been made that forest governance consequently provides one of the most useful lenses through which to scrutinise 'the increasing tendency for collaboration in many sectors where political and economic trade-offs also exist'.[3]

But as these observations imply, forest governance is an evolving project, and calls for some historical and theoretical investigation before it can be understood in terms of its contemporary institutional manifestations. The first part of this chapter therefore prepares the ground for the following case studies by looking at some of the central developments in forest governance in the decades before and after the Rio 'Earth' Summit. The second half takes stock of these developments from a theoretical perspective, building on the organising concepts presented in the Introduction. A short conclusion summarises these discussions, and points towards the institutional analysis that follows.

Forest governance and the environment 1980–2000

Intergovernmental initiatives to combat deforestation

In 1983 the UN General Assembly established The World Commission on Environment and Development (WCED). Chaired by Gro Harlem Brundtland, the Prime Minister of Norway, WCED was given the task of developing an agenda for environmental change at the global level by the year 2000, and beyond. This was to be built upon what has now become the foundational concept of contemporary geopolitical action on the environment: sustainable development.[4] The 'Brundtland report', published in 1987 under the title *Our Common Future*, was highly influential in framing the debate about the growing environmental crisis, particularly at the intergovernmental level. In 1992, the UN General Assembly, influenced by the report, which urged that its conclusions be turned into a UN Programme of Action for Sustainable Development, determined to convene the UNCED. Various preparatory discussions were held to develop conventions on climate change, biological diversity and a possible forest convention; to look at adopting measures to control land-based pollution and the production of an Earth Charter to set out principles of conduct for environmental protection and sustainable development; and finally, to adopt a programme to implement these principles to be called Agenda 21. In the end, intergovernmental discussions broke down into now-familiar camps. The global South accused the industrialised nations of the global North of trying to undermine the sovereign right of the South to exploit its own natural resources; the industrialised nations were also accused of attempting to externalise the costs of the environmental problems that the developed world had first created onto the developing states. As a result, the Rio conference was not asked to adopt an Earth Charter, or conventions on land-based pollution or deforestation. However, a watered-down version of Agenda 21 and a Declaration on Environment and Development, as well as conventions on climate change and biological diversity were adopted.[5]

Nevertheless, UNCED played an important normative role in shaping the general response to the environmental crisis. The evolution of global environmental governance can be historically linked to the growth of non-state participation within the UN and its institutional structures. Two significant developments resulted from UNCED. Firstly, major corporations (such as ICI) as well as private philanthropic organisations (including the MacArthur and Rockefeller foundations) supported and sponsored the event. Secondly, civic interests played a role in the negotiations in the lead-up to UNCED, and were involved in a number of preparatory committees. This is to be contrasted with the earlier UNCHE, where civil society participation was confined to the identification of items for discussion. As a consequence of these developments, large numbers of non-state observers were present at the Rio conference to lobby government delegations.[6] The extent of non-governmental organisation (NGO) participation in intergovernmental activities around the environment enhanced the degree of recognition accorded to public participation by state interests. Increased governmental recognition of non-state interests also reflected the broader normative impact of the role that was played by NGOs in other intergovernmental decision-making processes, such as those relating to aid. However, within the environmental policy domain, it has also been argued that while it may appear that NGO concerns are reflected in the language of the negotiated texts arising from intergovernmental environmental processes, this should not necessarily be taken to demonstrate *actual* NGO influence.[7]

Swedish Prime Minister Ola Ulstein has been identified as first putting forward the proposal for a forest convention as a consequence of the inability of the existing international programmes to protect forests.[8] Nine separate proposals for a global forest initiative have been identified as taking place between the months of January and December 1990. The FAO's 'Possible Main Elements of an Instrument (Convention, Agreement, Protocol, Charter) for the Conservation and Development of the World's Forests' emerged as a major contender, but was ultimately eclipsed by the UNCED process itself, although some of its language reappeared in UNCED documentation.[9] UNCED and its successors have been particularly criticised by a number of scholars in the field of forest governance for failing to take effective action to combat deforestation. Despite the promises and optimism generated at Rio for a global approach to forest issues, none was reached, and subsequent forest-related agreements (including the 1994 renegotiated International Tropical Timber Agreement – ITTA), were much weaker than expected. Rio was ultimately unsuccessful in bringing to forests the same degree of cooperation as to climate change, biological diversity and desertification, and the widely anticipated legally binding instrument (LBI) on forests did not eventuate.[10]

Various intergovernmental institutions within the UN system were involved with the forest-related policy decisions arising from the substantive Rio document, *Agenda 21* (Chapter 11, combating deforestation), and the related *Statement of Forest Principles*. The *Forest Principles* employ much of the language first used in the documentation associated with the earlier FAO initiative.[11] As with other Rio materials and going beyond FAO, it included relatively strong language relating to non-state participation in forest-related policy-making. But despite these sentiments, Rio failed to deliver the same degree of cooperation on forests as it did with climate change, biological diversity and desertification. These issue areas resulted in a series of formal conventions, but the much-anticipated LBI on forests did not eventuate.[12] One of the very few points on which parties could agree was that National Forest Programmes (NFPs) should be continued. NFPs had been first mooted as a means of improving forest management during discussions between donor countries and international organisations on the development of the Tropical Forestry Action Plan (TFAP). They foundered there, however, when support was withdrawn by international organisations following evaluations revealing that NFPs were failing to deliver expected on-the-ground impacts. The idea re-emerged during UNCED negotiations, and was extended to cover all forest types.[13] Generally however, UNCED's inability to combat deforestation comprehensively has been identified as a catalyst for the growth of forest certification.[14]

On the national level the most significant forest management forums were the Helsinki (later pan-European) and Canadian Montreal processes, aimed at developing C&I against which the sustainability of forest management could be assessed.[15] Others include the Tarapoto Proposal, Dry Zone Africa Initiative, North Africa and Near East Initiative and the Central American Initiative of Leparterique.[16] The Europeans coordinated their efforts to develop their own C&I from 1993 onwards through the Ministerial Council for the Protection of Forests in Europe (MCPFE). Six criteria and 101 indicators for SFM were developed in the years 1994–1996, and an agreement was reached that these C&I should be further elaborated at the sub-national and forest management unit levels.[17] These were followed in 1998 by set of Pan-European Operational Level Guidelines (PEOLG). These were identified as a voluntary set of recommendations were designed to put international obligations into verbal form to describe how SFM could be put into practice, but which had no legally binding status as regulations. The Guidelines, it was stated, had the potential to form a voluntary reference for the development of standards, and were indicative only, but it was also stressed they should not be used to determine SFM in isolation.[18]

Through 1994 and early 1995 Canada successfully created an international framework for C&I for non-European temperate and boreal forests. At the outset, eight countries participated in the process (Australia, Canada, Chile, China, Japan, South Korea, Mexico, New Zealand, Russian Federation and the USA). By 1997 Montreal included 12 signatory countries making up 60 per cent of the world's forests and 45 per cent of the world trade in wood and wood products.[19] Under Montreal SFM was not defined, nor did it detail prescriptive measures that would result in SFM. This approach resulted in a definition of the C&I as simply representing 'a common understanding of what is meant by sustainable forest management... the C&I are tools for assessing national trends in forest conditions and management, and provide a common framework for describing, monitoring and evaluating *progress towards* sustainability at the country level.'[20] This approach was justified under the rationale that all criteria taken together contributed towards SFM.[21]

Shortly after the commencement of these initiatives, the Intergovernmental Panel on Forests (IPF), which functioned from 1995 to 1997, was formed as a forum for forest policy decisions.[22] Quite early in the IPF's policy deliberations it was determined that there should be a structural separation between C&I and the emerging processes of forest certification, and it was recognised that contrary to the intention of forest certification to accredit operational performance, C&I were not be used as performance standards for the certification of forest management.[23] In 1997 a successor body, the Intergovernmental Forum on Forests (IFF) was established, running until 2000. In terms of substantive outcomes, the IPF/IFF deliberations generated 270 Proposals for Action (PfAs). In 2000 a third body, the United Nations Forum on Forests (UNFF), was created. NFPs were to survive throughout the IPF/IFF and subsequent UNFF negotiations, and emerged as means of voluntary national-level reporting on forest management activities.[24] NFPs have been described as instruments to ensure more rational long-term planning and coordination of forest policy, which are designed to replace traditional technocratic approaches.[25] However, the value of such approaches has been questioned and all three of these processes have been dismissed as being based on lowest common denominator politics, producing sub-optimal outcomes.[26] They have been interpreted as part of a long line of competing, and/or concurrent – but ultimately failed – attempts to tackle deforestation.[27]

The birth of forest certification

Public concern about global deforestation became pronounced in the 1980s. The literature outlines a historical narrative in which NGOs became increasingly active in their attempts to influence global timber policies and processes. This was partly a response to calls from social and environmental groups in the global South for assistance in helping them to save their forests, and recognition of the failure of the international policy community to solve the problems of forest degradation and deforestation. NGOs in the North began to look at the tropical timber trade in particular, and ways to influence that trade. They began to place forest conservation campaigns within the context of trade, resulting in the creation of a range of trade-related strategies. These ultimately led to an expansion beyond tropical rainforests to include temperate and boreal forests and involved large numbers of NGOs carrying out many interrelated but separate campaigns: targeting particular kinds of businesses (such as retail stores selling timber, or transnational corporations associated with the international timber trade); trying to persuade local and state governments to refrain from using tropical timber in building projects; and pressurising national governments and supranational bodies, such as the EU, to ban tropical imports. By 1988 NGO efforts in Europe had been successful enough to bring about a vote in the EU Parliament, which recommended an import ban be put in place by EU members until logging of tropical rainforest – particularly in Malaysia – became sustainable, although this was later overturned by the EU Commission. By 1989, the call for a boycott had spread beyond Europe to groups in countries such as Australia, where activists began blockading ships, and in the US and Europe to a full-scale boycott of companies, implicated in tropical rainforest logging. By 1990 the impact of local government bans in Europe caused Malaysia and Indonesia to threaten trade retaliations. In 1992 the Austrian government imposed its own import ban but dropped it a year later after Switzerland, which looked like following suit, was threatened with a counter-ban against Nestlé by Malaysia. In 1993 Dutch NGOs signed an agreement with the Dutch government and timber importers, which established a 1995 deadline on the importation of unsustainable timber.[28]

While some NGOs pursued export bans and timber boycotts, others looked to existing intergovernmental initiatives and international trade regimes, such as the International Tropical Timber Organisation (ITTO), as potential levers for change. NGOs began to lobby ITTO members to impose sustainability requirements on the tropical timber trade.[29] The ITTO had been founded in 1983 by 64 countries, under the auspices of the United Nations Conference on Trade and Development (UNCTAD), and comprised producers and consumers of tropical timber who were given a mandate to consider global resource management issues. In 1985, the International Year of the Forest, a range of UN institutions continued to work on the development of a number of forest initiatives.[30] In the same year, the ITTA, considered to be the first commodity agreement designed to include forest conservation, and one of the chief products of the ITTO, received enough signatories to become ratified, but with a slow take up from producer countries it did not gain legal recognition until 1994.[31] Further initiatives included the finalisation of negotiations regarding a TFAP, which had commenced in the early 1980s and was concluded in 1986 with the Plan's endorsement at the eighth session of the FAO's Committee on Forestry (COFO). However, all these initiatives have been accused of failing to respond effectively to the dynamics of globalisation that produced such environmental problems as deforestation.[32] The ITTO in particular, it has been argued, was undermined by political compromise and as a result was simply unable to act decisively.[33]

In 1985, Friends of the Earth (FoE) in England and Wales proposed what has been claimed to be the first modern timber certification and labelling scheme as part of a campaign to save tropical rainforests.[34] The labelling of wood products has been traced back to a French royal decree of 1637, the purpose of which was to designate the quality of the products labelled, rather than quality of management, as was the intention here.[35] Consumers in Britain and Europe were encouraged to avoid purchasing tropical timber produced on a non-sustainable basis, and the organisation launched its own Good Wood programme. From 1988 onwards, stickers and tags began to be attached to wood products as a seal of approval.[36] These and other NGO initiatives focusing on tropical deforestation generated a reaction within ITTO. In 1988 it commissioned a report from the International Institute for Environment and Development (IIED), which found less than 1 per cent of the global timber trade to be from sustainable sources. This report encouraged both NGOs and the ITTO membership to look more closely at promoting SFM.[37] In 1989, FoE, possibly encouraged by the IIED report, prepared a proposal for the ITTO to look at the feasibility of developing a timber certification and labelling programme.[38] The ITTO Council received the proposal but rejected the recommendation.[39] It dismissed the proposal as 'a veiled attempt...to encourage the current campaign of boycott against the import of topical timber products'.[40] FoE's original concept was transformed into a broader initiative aimed instead at developing incentives for SFM generally.[41] The Council's action effectively placed any opportunities for certification via the ITTO in abeyance and as a result, FoE withdrew from the process.[42]

ITTO's decision only served to encourage NGOs to begin organising around certification more seriously. Increasing NGO disillusion over ITTO's lack of action in combating deforestation, and its rejection of forest certification, led another international environmental NGO, the World Wide Fund for Nature (WWF) to warn ITTO that if it failed to start working towards tropical forest conservation, WWF too would look at alternative means to achieve this.[43] The Council's anti-certification decision consequently also acted as a catalyst for WWF to create a programme aimed at moving the tropical timber trade into sustainable sources by 1995. This resulted in a major shift in its campaign focus away from intergovernmental- and towards non-governmental private sector initiatives.[44] By the turn of the decade, more NGOs players had come onto the scene, and the Rainforest Alliance, a US-based non-profit organisation, formed the SmartWood Program, constituted for the purpose of 'forestry management certification'.[45] The programme's first certification was carried out in November 1990 in Indonesia, covering the teak wood forests of Java, managed by Perum Perhutani, the state forestry agency.[46] In 1991, WWF commenced discussions with business and industry in the UK and began to establish trade networks, designed to facilitate the transition of timber-consuming businesses to sustainable timber sources.[47]

These were the background events that generated the impetus for the foundation of the FSC. WWF directly attributed the impetus to move forward on certification to ITTO's lack of action, and predicted that the concept would 'leave the ITTO behind'.[48] Part of the problem with NGO relations with ITTO was that NGOs and trade-related interests placed different emphases on the conservation/development objectives implicit in the concept of SFM. These differences were ultimately to be expressed via competitive industry- and NGO-driven market-based systems of forest certification.[49] It has been argued that NGO/industry competition originated from the fact that NGOs were simply responding to the consuming public's scepticism over the validity of claims made by 'self-certifying' individual companies, and the need for arm's length approaches. Governments' closeness to business interests was seen as resulting in weak measures, making it inevitable that civil society groups would enter the certification market in earnest.[50] Non-governmental forest certification (as opposed to the industry-sponsored, and government-supported initiatives discussed above) has been characterised as an alternative system of forest management that created its own principles and criteria (P&C) independently of governments.[51]

After the advent of forest certification, the mass consumer movements of the era of the tropical timber boycott increasingly lost momentum, but not without conflict within the NGO sector. On the one hand, major NGOs, exemplified by those involved in the development of certification, favoured voluntary, private initiatives as a means to encourage SFM. This was consistent with the tenets of liberal environmentalism, which was gaining ground at the time, but was contrary to previous environmentalist efforts to have the forest industry formally regulated.[52] Smaller national, and more 'activist' groups objected to the commodification and corporatisation of forests and they believed forest certification would be co-opted by economic interests.[53] Consequently, the trade-related strategy came into conflict with national environmental groups still pushing for a blanket ban on purchases from tropical countries.[54] But these were gradually superseded by a greater campaign emphasis on developing demand for certified timber amongst timber traders. A Global Forest and Trade Network was established by WWF in 1997, servicing buyers' groups, which had been established in the UK, the Netherlands, Belgium and Austria.[55] Campaigning in this direction reached a high point in April 1998, when WWF signed an agreement with the World Bank, resulting in the WWF-World Bank Forest Alliance, under which 200 million hectares of production forest were to be independently certified by 2005.[56] To a certain extent the concerns of the 'pro-intervention' NGOs were borne out by the co-option of government-led initiatives by industry, and the instigation of industry-based certification.

The rise of corporate environmental standards

While Rio was less effective than anticipated in bringing state-based regulatory action on the environment, it played a central role in the growth of the standards-based approach to environmental problem-solving. The claim is made that the standards that emerged in the post-Rio environment represent a move away from attempts to govern the environmental and social excesses of corporate behaviour by more aggressive regulatory means. Rio has been seen as paving the way for voluntary and self-regulatory initiatives, often developed directly by corporate interests. Negotiations to develop a corporate code of conduct, including environmental provisions, had been under way since 1977 under the auspices of the UN Centre for Transnational Corporations (UNCTC). These were abandoned shortly before Rio, under pressure, NGOs claimed, from the US and the International Chamber of Commerce (ICC). This left the way open for voluntary initiatives whereby corporations developed their own self-regulatory measures.[57]

Environmental management standards reflect the thinking of economics scholars in the late 1980s that pollution essentially represented resource wastage, encouraging a perspective that pollution control was to be seen as a matter of quality assurance.[58] Early initiatives by the industrial sector in response to public criticism, such as the chemical manufacturers' Responsible Care programme, as well as the creation of national environmental management standards, exemplified by British Standard 7750, can be traced back to this philosophy.[59] In the years immediately prior to Rio, and following the widespread and successful piloting of British Standard 7750, the EU developed an Environmental Management and Audit Scheme (EMAS). This scheme was voluntary, but did require participating industries to make a commitment to continually improving their environmental performance and subjecting themselves to an external audit by an accredited verifier if they wished to be certified.[60]

In 1991, the ICC and the Business Council for Sustainable Development (now the World Business Council for Sustainable Development – WBCSD), moving beyond the European sphere, approached the world standards-setting body, International Organisation for Standardisation (ISO), to develop an EMS.[61] The Rio organisers also made similar requests.[62] ISO's response was to establish the Strategic Advisory Group for the Environment (SAGE). SAGE's work on environmental management set the technical tone for the subsequent lobbying and content development that occurred during the UNCED preparatory conference in January 1992. The group's recommendations ultimately became some of the core principles underlying the substantive documents of UNCED, and contributed to the comprehensive policy guidance contained in *Agenda 21* as well as the Rio Declaration itself.[63] Corporate responsibility in this context was depicted as being complementary to the existing environmental regulatory and legislative arrangements of the state. Operational standards were not required to specify absolute environmental performance requirements, but were based upon a more limited set of obligations, whereby environmental commitments were confined to providing a framework under which a company determined its own internal environmental management priorities, and systematised them accordingly.[64]

It is also important to note the global trade environment in which both the ISO EMS and the EU EMAS schemes were developed, especially the influence of the Uruguay Round of negotiations and the creation of the World Trade Organisation. Here, voluntary *process* (as opposed to *product*) standards were encouraged as a means of addressing environmental concerns without violating trade rules. Both schemes have been portrayed as seeking to pre-empt the proliferation of national environmental laws, which, it was seen, could act as barriers to trade.[65] There are substantive differences in the reporting requirements in the two systems, and the strength of language used. ISO 14000 was – and continues to be – less explicit about exactly what degree of 'continual improvement' was expected in environmental performance, and unlike EMAS did not originally require participants in the scheme to publish an environmental statement.[66] ISO has emerged as the undisputed global leader in environmental management, and its outputs are recognised as the trade-legal standards under the WTO's Technical Barriers to Trade (TBT) Agreement. It consequently wields significant influence over business, governments and civil society at the global level.[67] EMAS is based on national and regional initiatives, which makes it less applicable at an international – as opposed to European – level.[68] ISO's pre-eminent global position has exposed it to an intense discussion about governance, particularly in matters relating to stakeholder participation.[69]

Soon after Rio the forest industry began to see the value of adopting standards-based initiatives as a means to counter emerging NGO schemes. These were first expressed in the market through certification schemes in North America, where forestry interests responded to the creation of the FSC in 1993 by commencing the development of their own competitor certification systems: the Canadian Standards Association (CSA) and the US Sustainable Forestry Initiative (SFI). CSA, in an alliance with Australia and other industry interests within the ISO family, also tried to insert their vision into the ISO 14001 certification programme, but this proved unsuccessful. The reason for such a strategy, NGOs argued, was to avoid the environmental stringency that labels such as the FSC would require.[70]

This failure made the forest industry pay closer attention to C&I processes occurring in Canada and Europe at the time. The forest industry was not naturally supportive of such measures, but its use of C&I can be traced to a realisation by the mid to late 1990s of the necessity of developing an alternative market programme. While Canada pursued its own national C&I-based project, European efforts were led by Scandinavia, and between 1995 and 1996 forest industries and forest owners in Norway, Sweden and Finland attempted to develop a 'pan-Nordic' project. This foundered on account of NGO opposition to anything other than FSC certification. A broader alliance of European forestry interests, by this time including France and Germany, consequently gathered around the Helsinki C&I and the PEOLG, adopting them as the basis for their own programme, Pan-European Forest Certification (PEFC).[71] This was controversial, since the C&I were not designed as the basis for a standards-setting system, and MCPFE had not been prepared to formally endorse any certification scheme.[72] A link can therefore be established between C&I, NFPs and industry certification programmes. This co-option of pre-existing processes has been interpreted as injecting confusion into the forest certification debate.[73]

Theorising contemporary forest governance

Describing the phenomenon

The analysis of the historical developments in forest governance discussed above is not confined to any one theoretical school, and is subject to varying interpretations. International forest deliberations of the intergovernmental variety, although challenged regarding their effectiveness, have not been wholly replaced, and therefore sit alongside and complement a number of newer governance types, including NFPs, forest certification, and various other public, civic and private initiatives.[74] This has resulted in some confusion in the literature as to whether these separate initiatives represent forest regimes in their own right, or whether they should be treated as discrete parts of an international forest regime. Humphreys notes the disputes within the literature questioning the existence of an international forest regime, since there is no international forest convention (although as of 2008 there is a non-legally binding instrument – NLBI). He agrees with European forest policy scholars that such a regime emerged during the mid 1990s and argues for a revised definition of a regime as encompassing three aspects: the more commonly recognised 'hard' law arrangement of a single international legal convention backed up by subsequent protocols; 'soft' law, such as NLBIs increasingly preferred by states, including the resolutions adopted by UNFF; and what he also refers to as private international law.[75] Such a three-fold analysis may unduly restrict the regime concept, and it may be too early to identify any normative regime consensus around which actor expectations are converging. There is general agreement amongst scholars, however, that contemporary forest governance reflects the trend away from top-down approaches towards new governance, notably governance by networks, of which forest certification has been categorised as a classic example.

Certification, it has been argued, has had a more powerful influence on forest owners than government initiatives and has proved more successful in terms of protecting areas than public approaches, such as nature conservation agreements. It has been identified as consisting of a form of private governance with government, having the potential to complement, but not replace, public instruments, particularly since governments remain a key actor in both systems. Certification plays an important role in terms of government policy implementation because it not only affects the behaviour of certified forest owners but also influences large numbers of other forest owners. The willingness and capability of such groups to assimilate information and correct their behaviour are essential for successful policy implementation.[76]

Cashore et al. present another perspective, arguing that certification is to be understood as a new institutional form of sustainable development beyond existing governmental processes. Such private governance systems may improve environmental performance across the board in ways that traditional public, command and compliance models have not. The resulting voluntary-compliance market mechanisms coincide with increased civil society demands at a time of reduced government spending, and has created a form of international 'liberal environmentalism' that seeks to avoid command-and-control responses and avoid the traditional business-versus-environment dilemma. However, the designation of all such types of governance as 'private' is over-simplistic, as it puts voluntary codes, public reporting and certification together when they are in fact very different governance mechanisms. The problem with such an approach, it is argued, is that it ignores the rapid development of systems of international authority, which are not driven by the state. The loss of state authority, including the granting of legitimacy to alternative venues of power – particularly market-based instruments – is worthy of examination in its own right. Forest certification has been located within a general movement that displays a trend towards international private/civic governance, in which actors largely govern their own affairs, without state involvement, or without the legitimisation of state political authority. This has led to the definition of forest certification, as discussed by Cashore et al. as an example of non-state market-driven governance (NSMD). National forest certification, exemplified by PEFC, has both influenced – and been influenced by – the changing laws and policies associated with SFM and NFPs. This has led to a revision of the classification of certification, which is now seen as operating in the shadow of governmental hierarchy, and an example of the 'mixed' mode of governance.[77]

Forest certification still requires compliance with government regulation. Governments can also act as stakeholders, procurers and users of certification, and can provide resources to assist those seeking certification, as well as participate in standards development. They can also determine which schemes to support, and which not. This support might be financial, or symbolic. However, the state does not exercise sovereign authority in requiring adherence to NSMD governance systems. Rather, a whole range of organisations makes decisions as to whether to support such schemes. Participants are generally the same as in public policy-making (environmental NGOs, business groups, professional and trade associations) and they act as representatives for the broader public, who grant them their authority based on their shared values. Government simply becomes an interest group. This has impacted on traditional notions of authority, which now sit alongside non-state and shared private–public concepts (see Table 2.1 below).

Whilst it is not yet clear if non-state systems will complement or challenge the nation-state, or if the nation-state will seek to simply absorb the phenomenon in some way, non-state systems mark a radical departure from the traditional Westphalian sovereign authority model of public policy. As a new governance phenomenon NSMD systems are seen as constituting a new experiment in how procedures and rules are established, which departs from traditional methods; in this regard alone, certification can be considered significant.[78]

The idea that forest certification has something to offer governance as a whole, is elaborated by other commentators. Meidinger considers forest certification to represent a new form of administrative law. Its procedures, due to their dynamic nature, encourage learning amongst participants; the rapid flows of information and multi-stakeholder dialogue (MSD) and debate are valuable contributing mechanisms, but as it is still a new phenomenon, he is unsure if genuinely beneficial, as opposed to opportunistic, manoeuvring is occurring.[79]

Table 2.1 Comparison of non-state market-driven governance sources of authority with other forms of governance

Feature	Non-state market-driven governance	Shared private–public governance	Traditional government
Location of authority	Market transactions	Government gives ultimate authority (explicit or implicit)	Government
Source of authority	Evaluations by external audiences, including those it seeks to regulate	Government's monopoly on legitimate use of force, social contract	Government's monopoly on legitimate use of force, social contract
Role of government	Acts as one interest group, landowner (indirect potential facilitator or debilitator)	Shares policy-making authority	Has policy-making authority

Source: Cashore et al. (2004, p. 28).

Quality of governance and definitional inconsistencies

Forest governance theorists acknowledge that 'new' governance has resulted in changed conditions for demonstrating effectiveness, and that institutional performance is understood in terms of input and output legitimacy.[80] Effectiveness describes overall performance in relation to an institution's objectives or programmes, and efficiency denotes the cost and rate by which inputs translate into outputs.[81] Meeting the needs of the diverse stakeholder groups inherent in such systems can create internal tensions and impact on notions of legitimacy amongst stakeholders. Different strategies have been undertaken to achieve legitimacy.[82] This has resulted in an analysis of the rivalry between forest certification schemes in particular as being understood in terms of the different approaches adopted within each system to the common problem of gaining rule-making legitimacy between NGOs, forest owners and forest product purchasers.[83] Looking at the competition between industry schemes and the FSC, some scholars conclude that it is less the type of legitimacy that is significant, but the strategies undertaken to achieve legitimacy. Sources of legitimacy range from the purely pragmatic, which is narrow and self interested, to the moral, based on ethical values. Legitimacy can also be derived cognitively, meaning a given scheme is understandable, or simply taken for granted. Legitimacy in this latter context can occur if an institution has been in existence long enough to become a normative institution, or by mirroring existing institutional structures, such as the UN. Non-state forest governance systems in particular have been identified as drawing their legitimacy from their NGO base.[84]

Regardless of the analysis, certification has been identified as indicating a wider trend in the administrative law associated with global governance as it questions traditional notions of how political legitimacy is conceived within transnational regulation.[85] In forest certification, quality of governance and legitimacy have been directly linked to participation. Other forms of forest governance beyond certification have thrown up similar discussions on how to determine quality of governance. In the case of national forest policies, participation and deliberation are seen as significant constitutive elements of forest governance, contributing to greater collaboration and conflict resolution.[86] Interest representation, as a sub-set of participation, is also seen as an important aspect of quality of governance, particularly in certification. The logic underlying this is that a 'good' standard is legitimate because it represents the concerns of the parties that developed it. Although there are some problems with this theory, it is difficult to argue the legitimacy of a 'bad' standard that does not reflect the concerns of interested parties.[87]

Given the diverse nature of participants in forest governance, how these interests are balanced has consequently been identified as being crucial for broader credibility and rule-making legitimacy.[88] There is a special need to balance interests effectively through inclusiveness, which is seen as being instrumental in regulatory credibility and authority. Inclusiveness provides for a mix of qualities and resources amongst participants, allowing for the combination and mutual adjustment of interests. Participation in standards-setting by such diverse groups also appears to build trust and enhance common expectations and understandings, as well as allowing for a certain degree of consensus. A further need for power distribution to prevent any party from becoming dominant in the general regulatory space has also been identified. Returning to forest certification, there are examples of controversies between the FSC and its competitor PEFC where demands for inclusiveness have come into conflict with industrial interests over control of standards-setting.[89] Certification may have increased the breadth of participants in forest policy arena, but questions have been raised as to whether industry-based programmes – in contrast to NGO-driven programmes – have balanced participation; although they have accepted it in principle, participation in terms of interest representation is tightly controlled.[90]

Consequently, certification systems use a range of criteria to demonstrate legitimacy. Aspects identified include credibility, comprehensiveness, objectivity and measurability, reliability, independence, voluntarism, equality, acceptability, adaptability, cost-effectiveness, transparency, practicality and applicability.[91] There are also various methods for quantifying the legitimacy of governance between competing programmes.[92] The World Bank serves as a case in point. It has developed its own principles for what it refers to as 'acceptable' forest certification, requiring 'meaningful participation' in standards-setting and fair, transparent, independent, and conflict-of-interest free decision-making procedures.[93] NGO commentators emphasise participation, transparency, consistency and performance-based standards as minimum requirements.[94] Forest governance scholars have also inadvertently contributed to the problem by also making use of various sets of criteria in their evaluations of quality of governance. Quality of governance is equated to legitimacy, which is derived from representativeness, accountability, transparency, equality and scientific veracity. On this view, the legitimacy of SFI and CSA has been rated as low to medium.[95]

As a result, determining the quality and legitimacy of forest governance presents a problem for scholars, public interests and institutions alike. Discussion has arisen in the literature as to what light 'ideal' models of forest governance can shed on such concerns. Forest certification, for example, has been presented as a 'true type' governance system, on account of its non-state orientation regarding rule-making authority, standards-setting and compliance verification. The point should be made, however, that care should be exercised when discussing the differences between different certification programmes. PEFC and FSC should not be overly simplistically interpreted in terms of whether one system is more 'pure' (say in the case of FSC, since it is almost wholly non-state in its orientation), and the other 'diluted' (in the case of PEFC because it derives some of its rule-making authority from the state).[96] But should also be stressed that these two models of forest certification should not be classified as one type. There is a tendency in the literature to talk about 'forest certification' generically, but this is not particularly helpful. As the following case studies demonstrate, the orientation of these two schemes, and their locations in the governance typology outlined in the Introduction, are quite distinct. The argument as to which governance system is more ideal is better made in relation to the manner in which the governance arrangements discussed in the

Introduction are expressed. In the following representation, the conceptual governance framework of Figure 1.1 and the PC&I of Table 1.2 presented in the Introduction have been combined to create an impression as to how legitimacy would be derived within an institutional model of governance quality. This has the effect of both providing a more elaborate way of understanding 'participation as structure' and 'deliberation as process', as well as mapping the critical dimensions of a governance system onto the institutional attributes that the PC&I are seeking to evaluate (Figure 2.1).

It should be noted in this representation, that unlike the hierarchical framework of PC&I in the Introduction, implementation is related

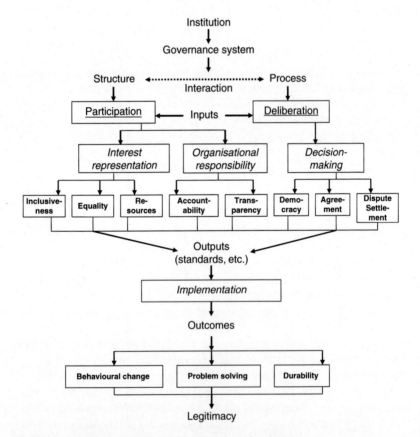

Figure 2.1   Ideal model of institutional governance quality.
Note: Boxed typeface indicates hierarchical relationship at the PRINCIPLE, *CRITERION* and **Indicator** levels.

to both structure and process, since both play a role in developing the outputs that need to be implemented. This difference arises from the attempt to understand governance from a viewpoint that takes a greater account of the complexity of the interactions within the system. The three criteria of interest representation, organisational responsibility and decision-making can be understood as being more directly related to input legitimacy, as it has been discussed above, since they concern issues regarding who is being asked to agree to the rules, and the democratic arrangements underpinning the system. Output legitimacy, being more concerned with the results of rule-making contributes more directly to the longer-term outcomes of behaviour change, problem-solving and durability. It now remains to be seen how the four institutions of forest governance selected match up to this ideal.

Conclusions

Governance has been presented as the primary means by which social and political interaction can be understood in the global context of the state, society and the market. This understanding has taken some time to eventuate, however, and the observation could be made that the sophistication of this understanding has evolved over the past two decades or so. And yet, concurrently, governance theory has directly affected practice, not merely in terms of retrospective analysis, but in shaping the strategic directions followed by various institutions. The historical period investigated in this chapter is characterised by three contemporaneous and interdependent sets of events, in which NGOs, governments and business engaged in an intense, and at times heated, struggle to define the contours of the institutional response to the global problem of deforestation. The governance systems that arose as a result of these interactions were both quite different and markedly similar. Initially, the attempts by NGOs to generate their own solution to deforestation by creating markets for 'good' wood were largely rejected by both business and government. But as the idea of sustainable development became normatively entrenched in the years immediately following UNCED business began to put greater efforts into developing its own market-based initiatives, in which it was – both wittingly and at times unwittingly – aided by a series of government processes happening at the same time. By the end of this period a range of institutions had emerged, which were collectively committed to solving the problem of deforestation by applying the concept of sustainability to the management of forests, but which went about pursuing this objective in different ways. This resulted in a wide range of institutional expressions, and inconsistencies between institutions.

From a governance theory perspective, this is not necessarily a problem, as it is unlikely that any single approach will suit the highly complex nature of these times. But it does raise questions as to how critical institutional arrangements like democracy function in such contexts, and how the rights, roles, and responsibilities of participants are both guaranteed and regulated. Determining performance is therefore not solely about institutional type, but is also concerned with investigating the extent to which global institutions meet the structural and procedural requirements of contemporary governance in the wake of Rio and the sustainability discourse. This is not to say institutional type is not significant, and it is important to determine the merits of one approach over another (intergovernmental models versus non-state systems, for example). But where there are differences in overall performance between two similar types, it is suggested that these may be more dependent on the extent to which different governance systems allow for participation as structure and deliberation as process. In the five chapters that follow the dynamic interplay between the actors involved in global governance is explored within a range of systems that function within different social and political settings. The nature of the interactions between these actors and these contexts forms the basis of determining the quality of governance, and ultimately, the legitimacy, of each of the institutions investigated.

3
Forest Stewardship Council

Historical overview

Origins, controversies and institutional development

At a meeting in San Francisco in 1991 a group of timber traders and social and environmental non-governmental organisations (ENGOs) agreed to develop an independently audited global system for 'good forest management', which, it was decided, should be called the FSC.[1] In March 1992 an interim Board was elected, with representation from both developed and developing countries and six working groups were created with a range of tasks including the drafting of principles of forest management.[2] Discussions during this period regarding the scope and intent of the emerging institution were wide-ranging. Initially it was proposed that FSC should be created as a foundation, with no members and only a Board of trustees, but the working group was persuaded to adopt a more participatory structure, which it was argued, would provide the FSC with greater legitimacy.

On account of what were seen as definitional controversies surrounding the term 'sustainably managed' it was decided instead to draft a series of P&C, the intent of which was to deliver 'well-managed forests'.[3] Although it was agreed that FSC should address all forest types, deliberations regarding the scope of FSC's forestry mandate were not easy. Northern NGOs were accused of initially wanting certification to relate solely to the tropics, and the expansion to include temperate and boreal forest may be attributed to concerns that a focus on the tropics in NGO campaigns had caused consumers to regard timber from such areas as inherently 'bad', whilst non-tropical timber was 'good'. National NGOs in timber-producing countries in Scandinavia and North America especially disputed the assertion that forest management in their countries was sustainable. Conversely, some producers from the tropics sourced their timber responsibly, but were being denied the opportunity to market it as such. Fear amongst environmental non-governmental organisations (ENGOs) about losing credibility may have been another motivating factor, since the downturn in the tropical timber trade resulting from their campaigns had led to the clearance and conversion of forests to production of other commodities.

Consultation processes in ten countries also took place between 1990 and 1993 to determine the level of support for a global certification programme for natural forests as well as plantations. A founding assembly was held in Toronto in September 1993 consisting of 130 participants from 26 countries. Participants from developing nations were present and played a significant role, while indigenous peoples' representatives stated their case for the return of their lands. Forest industry representatives included the Canadian Pulp and Paper Association (CPPA) and Californian lumber and milling company, Collins Pine. A number of government representatives and public authorities attended as observers, while research foundations such as the World Resources Institute were also present.

Debate was intense as to whether economic interests should be allowed a vote in the proposed body, with Greenpeace and the WWF expressing reservations, whilst FoE argued most strongly that they should not. A proposal was floated that economic interests be allowed no more than 25 per cent of the voting power in the General Assembly, with the remaining 75 per cent being held by the social and environmental interests together. 13 social and environmental groups, including Greenpeace, FoE and indigenous groups withdrew their support, arguing that business had been given too much power, and they stayed only as observers. Discussions regarding the types of forest to be certified, including planted forests, also proved contentious, and the proposed principles of forest management were not finalised. The initially agreed P&C for natural forest management, of which there were nine, referred to plantations only in so far as their establishment was expressly forbidden on sites of environmental, cultural or social significance. One NGO eyewitness interviewed for this study characterised the meeting as follows:

> It was strangest meeting that I ever participated in. For two days and nights there was a running battle between the economic group and environmental and social stakeholders, who at that time were still

joined together. Both sides were dogmatic. The economic group said, 'We gave you all this money to bring you here.' And the environmental and social groups said, 'We don't trust industry. If you want FSC you will need to earn our trust. We don't want you to have any vote in the future board of FSC. You can have observer status, and you'll only get the right to vote once you deserve it.' There wasn't any preliminary agreement until just before the party on the last evening. A lot of the preparatory work had been done before the meeting, but throughout the event itself nobody believed the FSC was going to be established. It was only born at the very last minute.

A formalised 'chamber' system, in which social and environmental interests held 75 per cent of votes in one chamber, and forest owners and retailers held the remaining minority share of 25 per cent in another, was not developed until 1994. In the same year, the P&C were agreed upon and FSC was incorporated with its headquarters in Oaxaca, Mexico, consisting of a staff of three and an Executive Director, with a mission to promote 'environmentally appropriate, socially beneficial, and economically viable management of the world's forests'.[4]

The response to the development and founding of FSC was mixed. The reaction in some parts of the social, economic and environmental sectors that would be expected to favour FSC was not entirely supportive. Some social interests felt the founding assembly only superficially addressed their concerns about lack of consultation and discriminatory treatment of tropical countries. In North America, the major forestry companies responded quickly to the creation of FSC and moved to develop their own certification schemes. In Canada, the industry had not been especially interested in certification, but by the time of FSC Toronto meeting the CPPA (which sent a delegate) issued a press release announcing the development of a forest certification initiative under the Canadian Standards Association. In the US these efforts were matched by the American Forest and Paper Association (AFPA), which set about establishing the SFI. It was not only large-scale interests within the forest industry who were concerned about the arrival of FSC. European small forest owners met its decision to base itself in Mexico with suspicion. The move was interpreted as an attempt to appease southern interests, and there were objections that a body whose headquarters were so far away should interfere with the well-established traditions of forest management in Europe, which partly contributed to the creation of rival European scheme, PEFC.[5] In the tropics, Malaysia commenced the development of its own initiative in 1994. The Malaysian Timber Certification Council (MTCC) scheme presented its first draft national standard in 1996.

In the same year, and perhaps to extend its scope and appeal in the light of these developments, FSC softened its stance on the certification of plantations, and a revised set of P&C permitted the certification of plantations, with certain provisos.[6] This action may have appeased some forestry interests, but it was to cause growing concerns about FSC's credibility amongst some of those ENGOs that had withdrawn from the founding assembly in 1993. FSC also began to attract some criticism. At the beginning of 1996 its membership was drawn from only 25 countries and Asia and Africa were not well represented. Numbers in the economic chamber were still criticised as being too low to ensure proper interest representation. Under-representation of social interests, particularly from Africa and Asia was also identified as being a problem during this period, perhaps an even more severe one than economic participation. Small forest owners continued to accuse FSC of being discriminatory, in view of the fact that large-scale tropical forestry certification was both easier to achieve, and more cost-effective, than small-scale operations, and by 1997 only 4 per cent of FSC's area under certification was from small-scale or community operations. In 1998 the organisation instituted a 'group' certification programme, which allowed forest managers to organise collectively and by 2002 980,000 hectares of forests were certified under the initiative.

It has been argued that between 1993 and 1997 FSC and its supporters made a significant contribution to international forest certification, which might not have been as far-reaching without it. FSC provided an important forum for policy debate at this time, as well as stimulating the growth of competing initiatives. In 1993 an independent working group consisting of academics and NGOs developed an Indonesian eco-labelling standard (Lembaga Ekolabel Indonesia – LEI), which entered into cooperation with FSC in 1997. FSC's first accreditation contracts were signed with four certification bodies, and the first certified products bearing FSC logo were released in the UK in 1996. The first working group, also in the UK, was established in the same year to develop nationally relevant management standards. In 1997 the first national standard was endorsed for Sweden.

Serious concerns regarding FSC certification, and the manner in which it was implemented by its accredited certifiers, also began to emerge from the mid-1990s. One of the earliest, and most controversial, cases was the certification by the Rainforest Alliance's SmartWood

Program of teakwood plantations in Costa Rica. The operations were certified in 1995, prior to FSC's accreditation of its first four certifiers. In 1996 promotional advertisements were circulated claiming the plantations had been FSC certified. FSC initiated its own investigations about this possible breach of protocol, but declared that the use of its name had been unintentional and the certificate was endorsed under FSC programme in January 1998.[7] In 1999 NGO frustration over the failings within the FSC system were to culminate in the production of a highly critical report published by the Rainforest Foundation, *Trickery or Truth?* It challenged the commitment of certifiers to stringent – and consultative – certification assessments and the true extent of democracy, representation, transparency and what is referred to as multi-stakeholderism across the system.

Over the same period, the organisation consolidated a number of policy initiatives, which were brought together during 1999. FSC became the first certification programme to define high conservation value forests, and to delineate those forest characteristics that merited special protection. This was promoted by FSC as representing considerable progress in resolving forestry conflicts. Policy clarifications were also issued regarding the prohibition on genetically modified organisms from certified forests, the use of chemical pesticides, poorly defined in the P&C, and matters concerning contract labour. A major revision of the percentage-based claims policy, whereby certified and uncertified sources could be mixed, was also undertaken. This policy had been highly controversial amongst NGO members. These initiatives were instigated in recognition of the fact that many members did not feel FSC's policies had lived up to their expectations, and Executive Director Timothy Synnott accepted that mistakes had previously been made at all levels.

The 1999 General Assembly was marked by a number of governance-related motions, attempting to address some of the perceived shortcomings of the organisation. An experienced facilitator was used to guide the membership through the complex procedures now required for speaking to, amending, and agreeing on motions under a revised tripartite economic, social and environmental chamber system. Although these initiatives went some way to appease concerns, FSC was placed under increasing levels of scrutiny by NGOs, both supportive and sceptical. In May 2001, European NGO Forests and the European Union Resource Network (FERN) published its report *Behind the Logo*, a comprehensive analysis of FSC and its competitor schemes. It considered FSC to be the only scheme with sufficiently rigorous certification and accreditation procedures, and equitable decision-making. The report did note, however, that FSC was imperfect in a number of aspects, including methods of consultation, communication, and dispute settlement. Conflict had arisen largely on account of the practice of undertaking evaluations using 'interim' standards in the absence of national standards, a practice objected to by a number of NGOs. The Southern NGO World Rainforest Movement (WRM) also released several reports far less flattering than *Behind the Logo*, aimed at pressuring FSC to change its stance on a number of issues, particularly plantation certification. The main criticism arose from the negative impact the eco-labelling of monoculture tree plantations had on the struggles of local people in the global South to protect their social and environmental rights. In November 2001 the Rainforest Foundation published the report, *Trading in Credibility*, which systematically outlined a number of structural and procedural weaknesses in the FSC system, as well as some of the shortcomings of its certification and standards-setting activities in several countries. Interestingly, the validity of the report was challenged by Greenpeace, which claimed some of its assertions were overstated, and that many of the problems identified had already been addressed.

FSC's response to *Trading in Credibility* was relatively measured. The reception of the report in this manner may be partly attributed to the fact that during the course of 2001 FSC had already worked on a series of policies aimed at better addressing a range of specific – and controversial – issues. These included the development of a response to address concerns over plantation certification, the production of a draft chemicals and pesticides policy and the creation of a draft social strategy. Recognising that the organisation had both strengths and weaknesses, FSC began a process of decentralisation and transformation into a global network the following year. These activities included the placement of the Secretariat within an International Centre, which relocated to Bonn in 2003. During this period, regional offices were created for Europe, Latin America, Africa and Asia-Pacific. Over the course of 2002 work on building the capacity of the social sector and increasing FSC's responsiveness to and communication with its social constituents continued. Guidelines advising on how to incorporate International Labour Organisation (ILO) conventions into the P&C were also released. A group chain of custody policy, aimed at providing small timber processors with more options for accessing chain of custody (CoC) certification was also finalised. A related initiative to increase access to certification for small and low-intensity managed forests also commenced. At the forest management level a review of the organisation's plantation policies also commenced, while the chemical pesticides policy, following two years of 'intensive discussion and debate', was finalised.[8]

A significant number of motions debated at the third General Assembly were also related to governance and revisited some of the uncompleted business of 1999. In terms of institutional arrangements, perhaps the most symbolic decision made was to give equal representation to all three chambers on the Board, which was passed by an overwhelming majority. FSC decision-making in the General Assembly was also defined in more – almost microscopic – detail. Further measures to improve participation in policy development were also instituted. The percentage-based claims policy was further revised to allow highly mechanised sawmills dealing with only small quantities of wood to participate in the FSC system in an economically viable manner. But a motion to set the minimum volume of certified timber in solid wood products at 70 per cent was defeated. Motions to further encourage indigenous community-based access to certification, and to increase southern chambers' participation in the preparation and presentation of motions at General Assemblies, were also passed. Other motions aimed at improving the quality, consistency and transparency in relation to the activities of certifications bodies were also approved. Interestingly, an attempt to create a certification commission, and a new motion to investigate conflicts of interest between certification bodies and applicants for certification, were both defeated. It was further agreed that the use of interim certification using the certifiers' own generic standards, which had caused so much trouble amongst NGOs, would be phased out over time. This was in response to situations where a range of different plantation companies had been certified against inconsistent generic standards. Despite the work that had been done earlier in the year on the plantations policy, a majority of delegates (75.3 per cent) still considered the current draft to be unclear and it was resolved that further consultation was necessary to develop clear guidance on how Principle Ten should be interpreted. This was a timely reminder on the need to work further on plantation certification, as the WRM published another hostile report (cataloguing ongoing problems in Thailand and Brazil) in August 2003.

The years from 2003 onwards can be summarised as ones of further consolidation for FSC, following a major reorientation of the organisation into a decentralised global network. Several new or revised standards and procedures were released in 2004, along with a simplified trademark manual and labelling guide. Two of the four standards associated with the new CoC arrangements are worth mentioning. These related to the handling of non-FSC controlled wood from uncertified forests. The intention was to assist companies, governments and financial institutions to eliminate controversial sources of timber from their supply chains, particularly illegally harvested wood. The new standards were promoted as addressing the G8's action plan to combat illegal logging. They also represented, it was claimed, an option for governments who had begun implementing the action plan nationally by instituting procurement policies favouring legally harvested timber.

In 2004, a follow-up report to *Behind the Logo* published by FERN, *Footprints in the Forest*, compared FSC to the by now rapidly increasing number of competitor schemes. Although once again much kinder to FSC than most of the other schemes investigated, and still considering it to be the only credible scheme for most NGOs, the report was nevertheless critical of FSC on matters relating to both policy and performance. In particular, FSC needed to seriously address the problems that were associated with plantation certification. The continuous opposition to plantation certification was finally substantively addressed by FSC from 2004 onwards. In that year the organisation announced the launch of a review, issuing a discussion paper and holding an international meeting, where NGOs outlined several persistent criticisms. In February 2005 a policy working group was established, consisting of 12 participants – including the author – representing the three chambers each balanced North–South.[9] The working group presented its final report to the Board of Directors in November 2006. The report recommended one common set of integrated P&C, rather than retaining a separate Principle Ten for plantations. It was accepted that certification should be accessible to operations from across the spectrum of management, from low impact activities in native forests to intensively managed short rotation plantations; the higher the impact, the greater the emphasis on prevention, mitigation and compensation. Further recommendations of significance included putting measures in place to elevate the social components of FSC system to an equal footing with its economic and environmental aspects. A technical phase commenced in April 2007, with the guidelines and the revisions to the P&C presented to the General Assembly of 2008. A first draft of the revised P&C was made available for consultation in June 2008. This process was subsequently incorporated into a full review of the P&C (version 5.0).

In 2007, an expanded investigation in FSC's governance was undertaken. The review was partly in response to the Greenpeace publication *Holding the Line* (which challenged some of FSC's controversial certificates). A second, more formal phase commenced in February 2008 with the release of a white paper, *Options for FSC's Future*, aimed at FSC members and parties with a strong interest in governance, and a membership survey was conducted between June and July. Three substantive components were outlined in the review, relating to membership, international and national arrangements, and dispute settlement. The intention was to create a two-tier system of membership consisting of traditional FSC members and a new category of FSC supporters. A second set of proposals concerned the Board of Directors and senior staff (including national initiatives). The Board was to become more strategic and less operationally focussed, and would include non-voting technical advisors and a delegate from the national initiatives. A new layer of management, the Senior Executive Group, was also proposed, consisting of Directors from each of the FSC's business units, answerable to the Executive Director, to be renamed the Director General. Finally, it was proposed that an independent Disputes Resolution Committee be created, that the existing Board-based system of dispute resolution be disbanded, complaints procedures be streamlined, and an annual report detailing compliance actions be produced. The third phase of governance-related consultation commenced at the FSC General Assembly in November 2008, where the new proposals were discussed and universally supported, albeit with some amendments.

Institutional analysis governance within FSC

System participants

Both members and non-members may participate within the FSC system on the international, national and regional/local levels. Membership of the institution can be held by either individuals or legal entities (including governments under specific conditions) that support the association, its purposes and P&C. Formal enfranchised membership of FSC is broken down into three interest chambers. The members of these chambers are further sub-divided according to their country of origin, identified according to whether the country belongs to the global North or the global South. The economic chamber consists of those with a 'vested interest' in commercial forestry, either for-profit, or not-for-profit.[10] These include employees, consultants or representatives of forest-related companies, industry associations, wholesalers, retailers, certification bodies, traders, end-users and government owned or controlled companies. The environmental chamber is limited to non-profit NGOs, whose governing bodies are independent from governments and which are renewed by periodic elections or appointment. The social chamber consists of indigenous peoples' organisations and 'social movements'.[11] This includes non-profit NGOs and unions. Academic, research, legal and forest product associations are assigned to the relevant chamber, based on the nature of their activities. The membership application process is complex, and varies according to the sector applying. Membership fees, charged on a sliding scale on the basis of operating expenditure, but without discriminating against southern members, are levied to cover the costs of servicing the membership. There were 643 members of the organisation at the end of 2006. Members participate in the General Assembly according to the tripartite, chamber-based arrangement.

Institutional arrangements

International level

The structure of FSC is depicted in Figure 3.1 below.

The Director (now Director General) heads the organisation's Secretariat and is effectively chief executive officer. The Secretariat as a whole is responsible for supporting and guiding the day-to-day activities of FSC internationally. Located within the Secretariat is the Policy and Standards Unit, the role of which is to initiate the development of guiding documents, such as standards.

Historically, FSC has run its own accreditation programme 'to provide a credible assurance that they are competent and independent', and the programme is responsible for accrediting certification bodies according FSC standards.[12] It is also responsible for accrediting national initiatives (NIs) and FSC's own standards. FSC's original practice, whereby certifiers undertaking assessments were accredited under its own standards, generated accusations that the FSC was effectively its own accreditation body. In response, the Accreditation Business Unit (ABU) was created as an independent legal entity in 2002. In 2006 the organisation went one step further, establishing Accreditation Services International (ASI), a wholly separate legal body, which manages the FSC's Accreditation Programme as a for-profit business, with its own quality management system, and following the relevant International Standardisation Organisation/International Electrotechnical Commission (ISO/IEC) terms and definitions. As of June 2009 there were 22 accredited certification bodies within the FSC system.

National level

NIs are intended to assist in the promotion of FSC and make it more locally relevant and accessible. They have two main aims: to develop

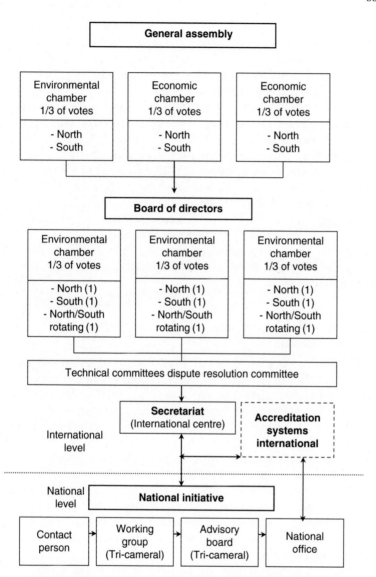

Figure 3.1 Institutional structure of the Forest Stewardship Council.
Source: Elliott (2000), p. 19 and Vallejo and Hauselmann (2004), p. 9, adapted. (Dashed lines around Accreditation Systems International indicate autonomy).

and test regional standards by encouraging local participation, with the assistance of the organisation's international membership; and to support the implementation and monitoring of certification activities. The expansion of NIs is built into FSC's annual business plan, which identifies those countries that are to be the focus of activities. There are currently 56 NIs within the FSC system and 30 approved national standards. NIs go through three developmental stages (contact person, working group, advisory board) until a national office is finally established. A wide-ranging set of participatory requirements is placed on NIs. These include dissemination of information to in-country members and other national interests, and updating the Secretariat. The extent to which information is circulated is determined by consultation with the Secretariat and the regional coordinator/director. Consultation is a major aspect of NIs, both in terms of the contribution of a national perspective on policies under development internationally and managing consultations with interests relating to certification activities nationally.

Standards development, accreditation, chain of custody and certification

FSC published a procedure for the development and approval of international standards in 2004. The procedures comply with the International Social and Environmental Accreditation and Labelling Alliance (ISEAL) Code of Good Practice for Setting Social and Environmental Standards (2004), ISO/IEC Guide 59 and the WTO TBT Agreement, Annex 3. Their development is intended to reflect transparency, participation and fairness by complying with international best practice. A number of FSC standards have been developed internationally. Standards development is relatively complex, but incorporates three main stages: initiation (request for the standard and formation of committees, staff and working groups); development (public consultation, drafting, testing, resolving disputes and approval); and implementation (publication, review and revision).

National working groups are responsible for the development of national standards. The purpose behind developing standards regionally is to make FSC standards locally applicable and workable. Locally defined forest management standards must be endorsed by FSC internationally, and are then used by certification bodies in their local activities. Standards development on the national or sub-national level is expected to demonstrate balanced participation and representation, accountability and transparency, and fair decision-making processes (including clear grievance procedures), and must harmonise with other regional standards. FSC forest standards are performance-based, defining levels of management that forest operations must achieve in order to be certified. Performance is based on the system's ten principles and 56 criteria, and although there is variation between national standards they must meet these P&C whilst still being appropriate to the country.

FSC accredited certification bodies may conduct both CoC and forest management certification for plantations and native forests, subject to the scope of their accreditation. They may also issue sub-licenses for the use of FSC name and trademark. The certification process is initiated through the request of a forest owner or manager. Where there is an adopted national standard it is used by the certifier as the basis for assessment. Where there is no endorsed national standard the certification body uses their own 'generic' standard, although this process is now being updated in line with recent General Assembly decisions.

Consultation of interested parties by certification bodies is required under Principle Four. A standard for consultation undertaken by certification bodies during their evaluation of a forest manager's activities was published in 2004. The extent of consultation is related to the scale of the operation under evaluation. Certifiers must have procedures in place for contacting interested parties, providing them with information, and hearing their points of view – if necessary, in confidence. In the case of the assessment of small or low intensity managed forests, certification bodies must carry out direct consultations with local and national parties where tensions are known to exist. Many of the certification bodies operate through affiliated organisations in various countries; these are not listed by FSC, but with the parent company.

Institutional typology

FSC has been depicted as being 'built out of an amalgam of sustainable development discourse, participatory, multi-stakeholder processes, technical standards-setting conventions, and emerging international trade rules'.[13] Its governance in particular has been characterised as consisting of delegated global transnationalism and in more recent, FSC-specific studies, as a form of global democratic corporatism. FSC is highly non-state since it was created largely by civil society (and business). This places it relatively far along the non-state end of the authority continuum. However, governments may play a role in the institution, although this role is relatively confined, and such interests are located

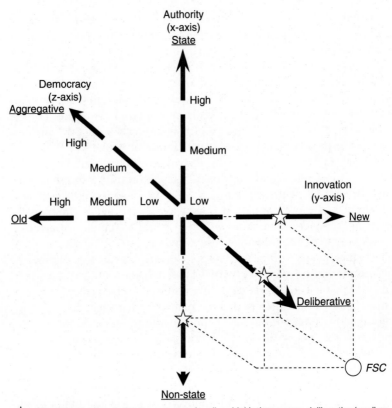

Figure 3.2 Institutional classification of FSC.

only in the economic chamber. Its authority is therefore somewhere between **medium** and **high** (see Figure 3.2 above).

FSC's democratic style is closely identified with deliberative approaches. The use of consensus decision-making is also the preferred mode of reaching agreement, although voting may be used when consensus cannot be reached. Voting here, however, is generally of a qualified majority nature, as discussed below.[14] This places FSC relatively far along the deliberative end of the democracy continuum, but taking into consideration the fact that voting occurs (even if it goes beyond the traditionally more aggregative system of the simple majority) it is again located somewhere between **medium** and **high**.

FSC can be seen as following global institutional norms by adopting the terms and structures of the UN framework. Here, the FSC's use of the General Assembly model, (even if it has been adapted) and the co-option of various UN normative definitions (such as those relating to geographical regions, and levels of national economic development or donor country status) also make it relatively 'traditional' in terms of modern global institutions. However, the system's participatory structure, which balances traditionally oppositional interests through its chamber system, and then further divides these sectors into northern and southern sub-chambers, is unique and goes far beyond the UN model. This sets it apart form other global governance systems; FSC clearly belongs on the 'new' end of the innovation continuum, and is once again located somewhere between **medium** and **high**.

Governance quality of FSC

Commentary

With a total of 39 points out of a maximum total of 55, or 71 per cent, FSC passed the 50 per cent threshold overall. One indicator received a very high rating (agreement), four high (inclusiveness, transparency, democracy and durability) six medium (equality, resources, accountability, dispute settlement, transparency, behaviour change and problem-solving), and no low or very low ratings. The conventional pass/fail target of 50 per cent was exceeded by all criteria (interest representation and implementation receiving 10 points or 67 per cent, organisational responsibility 7 points or 70 per cent, and decision-making 12 points or 80 per cent). At the principle level, the aggregate result for meaningful participation was 17, or 68 per cent, exceeding the target value of 50 per cent. The aggregate result for productive deliberation was 22 points or 73 per cent, also exceeding the target value of 50 per cent. The justification for these results is set out in the critical analysis below. See Table 3.1 below.

Evaluation

Interest representation

Inclusiveness

FSC has been identified as the most inclusive of the forest certification schemes in terms of the participation of a broad range of civil society participants, but less so for state actors. The programme's deliberate policy in its formative years of excluding governments from participating

Table 3.1 Evaluative matrix of FSC governance quality

Principle	1. Meaningful Participation					
Criterion	1. Interest representation Highest possible score: 15 Lowest possible score: 3 Actual score: 10			2. Organisational responsibility Highest possible score: 10 Lowest possible score: 2 Actual score: 7		Sub-total (out of 25): 17
Indicator	Inclusiveness	Equality	Resources	Accountability	Transparency	
Very high						
High	4				4	
Medium		3	3	3		
Low						
Very low						

Principle	2. Productive deliberation					
Criterion	3. Decision-making Highest possible score: 15 Lowest possible score: 3 Actual score: 12			4. Implementation Highest possible score: 15 Lowest possible score: 3 Actual score: 10		Sub-total (out of 30): 22
Indicator	Democracy	Agreement	Dispute settlement	Behaviour change	Problem-solving	Durability
Very high		5				
High	4					4
Medium			3	3	3	
Low						
Very low						
Total (Out of 55)						39

in the system on account of not wishing to anger NGOs (who felt state interests already had enough control over forest management) has been criticised. Following a decision at the 1999 General Assembly to investigate the participation of government bodies as members of FSC, a motion was passed at the General Assembly in 2002, which permitted government owned or controlled companies to join the economic chamber. Other government representatives are currently not eligible for membership, although the option exists for the creation of a chamber for government at the national level.

FSC standards acknowledge the importance of including national and local interests in the certification process. Both governments and NGOs are recognised as interests on the national level, as well as individuals and members of the community directly affected by forest management at the local level. Participants include any in-country FSC NI, state forest agencies and statutory bodies, national and regional ENGOs, indigenous peoples' organisations and forest dwelling- or forest using communities, labour organisations and forest workers, contractors and international NGOs active in a given region. Some businesses remain dissatisfied with the accommodation of their needs within the system. FSC's inclusiveness has been identified as a constraint in the effectiveness of its dealings with business.[15]

Interviewees expressed a range of views on the extent of inclusiveness within FSC. NGOs felt that although FSC gave local communities the power to discuss, influence, and work with the big firms around them, social interests had still been left somewhat behind. Business interests felt that FSC was too NGO-driven. Interviewees from the third group ranged from sharing the business perspective on NGO involvement, to considering FSC to be inclusive of private forest interests. However, private forest owners were confined to the economic chamber, despite having a strong social interest, to which some interviewees objected. Another argued that governments deserved representation in their own right in a single chamber. Finally, several interviewees across the groups saw a tension within the FSC system between active participation (in which participants had a very active role in standards development) and passive consultation (undertaken by certifiers purely for data collection purposes).

Considerable thought has gone into determining how participants (both members and non-members) should be included within FSC's governance structures from the international, through to the local level. These considerations would initially rank FSC's inclusiveness as very high. However, there have been, and continue to be, some problem areas. There is some tension among economic interests regarding the extent of their inclusion, and some weaknesses persist regarding social inclusion particularly at the local level. The same can be said for consultative processes relating to certification assessments at the local level. With these caveats, the inclusiveness of FSC is rated as **high**.

Equality

Interests who are not members of FSC have equal access to participation in FSC's standards-setting activities as members. Internationally, the development of standards encourages all interests and extra provisions exist for including the economic South. Non-members may also participate in technical working groups responsible for drafting standards. FSC's tripartite governing structure has been identified as being an innovative institutional model, which has prevented the organisation from being captured by economic interests and has provided a minimal standard for equal North–South representation.[16] The North–South imbalance in economic representation has been attributed to the increased compliance costs associated with ecologically complex tropical forest certification, and the prevalence of better regulatory standards in the North, which makes compliance comparatively easier, and certification more attractive. Small forest enterprises are in a minority, although improvements have been made in recent years. Economic interests were initially given less influence in the system, but in a pragmatic decision to gain greater business support the membership structure was modified in June 1996. Business interests play their role, but in a more limited fashion than in other international forums, defining FSC's specific institutional character.

The role of social interests in the FSC system has been acknowledged as weak in comparison with that of other sectors. This may be partly attributable to the historical fact that social interests in the FSC system did not obtain their own chamber until 1996. A social strategy published in 2002 acknowledged that workers, local communities, small forest enterprises and indigenous people in particular were 'marginalised politically' within FSC.[17] The strategy acknowledged in particular that this was the case for FSC certification processes. Equality of participation at the local level during certification assessments still remains limited. Matters raised by local interests impact the certification decision 'only in so far as its provides evidence of compliance or non-compliance with the requirements of the applicable FSC Standard'.[18] One response arising out of the strategy was to recommend more comprehensive consultation requirements for certification bodies to better assess social issues.

Information published in 2003 showed that developed countries held 66 per cent of all certificates and 80 per cent of the certified forest area. Industrial forestry interests also predominated, holding 35 per cent of certificates covering 66 per cent of the certified area. Most of these certificates were over 10,000 hectares. Community businesses held 25 per cent of certificates, but over only 3 per cent of the certified area. Boreal and temperate forests dominated over tropical and sub-tropical forests, with an increasing predominance of plantations. This imbalance in interests was exacerbated by the fact that certification bodies were also based in the North – even if FSC was taking steps to address these issues.[19] Numerically, the economic North sub-chamber has consistently dominated, with the social South sub-chamber being least represented. For the first time in 2006 the environmental South sub-chamber overtook its northern counterpart in terms of membership numbers.

The biggest single topic of debate amongst interviewees concerning equality was the influence wielded by WWF and other major ENGOs. WWF or other powerful groups often led NIs, for example. NGOs were also concerned that different interests were not always weighted equally during the certification assessment process, such as local communities. It was also hard for such interests to penetrate the FSC system itself if they were not part of the 'inner circle'. Although some interviewees qualified their criticism by stating that no one interest could dominate the system, others felt there was less room for other interests, particularly private forest owners. A number of interviewees explained that this was the reason why other certification systems arose.

FSC has put considerable effort into ensuring its participatory structures provide for equality, especially between members and non-members regarding standards-setting, but there is still a degree of asymmetry betweens northern and southern interests. Developing, tropical, small-scale interests do not benefit from the system as much as large-scale companies from the developed, northern/temperate zones. Certain specific interests are also 'more equal than others' within FSC system (most notably WWF). On these grounds, the score is **medium**.

Resources

During its establishment, much of FSC's funding was derived from private philanthropic institutions and government agencies. Historically its revenue has been derived through the increased commercialisation of services such as accreditation and trademark licensing, and over the 10 year period 1994–2004 this rose to a level of approximately 30 per cent to the organisation's global annual budget. Accreditation fees are levied according to the scale and location of the operation, and the extent of its conservation set-aside areas. According to one interview subject, up to 60 per cent of income has been derived from private philanthropic donors and government agencies in recent years.

The organisation provides resources for participation in a number of activities. Where a member is unable to attend the General Assembly due to economic constraints, they are entitled to seek financial support. NIs are expected to resource all aspects of their operations themselves. Interests consulted as part of the certification assessment process are generally not resourced.

Although FSC does provide resources for those in need, all those subjects interviewed – whether from developed or developing countries – commented that their involvement had been either self-funded, or their organisation had paid for their participation in the programme's activities. Both business and NGO interests alluded to the time involved, at both the personal and organisational levels. Costs could be considerable, ranging from employee time in several instances, to the funding of an entire NI as one NGO representative reported. One business interviewee wanted to stress, however, that sustainability certification was expensive in general. Their company had actually reduced expenditure by gaining ISO 14001 certification, and costs associated with FSC had been offset using the same management system.

The provision of funds to attend General Assemblies is to be highly commended, but participation in FSC's regional, national and subnational structures is less well resourced and varies depending on location. At the local level, where certification assessments and consultation of locally affected parties occurs, little or no resource support is provided for participants. Across the system, the provision of resources for interests to be represented is **medium**.

Organisational responsibility

Accountability

In forest certification there is an expectation that certification assessment is undertaken by a third party independent of the seller and buyer; hence the term 'third party certification'. Certifiers themselves need be independently accredited. Responsibility to ensure that the conditions of its accreditation are met is on the certification body. Any noncompliance during this period can result in the issuing of a corrective action request (CAR), which must be addressed within a given timeframe, or the Board can either fine the offending party or withdraw accreditation. Certificates are also issued for five years. Similar sanctions apply to the certified entity. Here, compliance is with FSC's P&C, and major failures will lead to de-certification. Certifiers were subjected to sustained criticism in the early years of FSC's development. Standards now exist requiring certification bodies to identify any issues that are 'difficult or controversial' in a region where a certification assessment is to take place.[20] They are required to consult with interested parties prior to the certification audit as to how the issues should be addressed in the evaluation. Certifiers are held accountable for their activities through compliance with the provisions laid down in the relevant standards, and certifiers have been suspended.

The unduly close relationship between FSC, accreditation and certification bodies, and standards-setting was a recurring theme amongst all three groups of interviewees. Far too much discretion was left to the certification bodies, which weakened the system considerably, especially when standards-setting bodies were also certifiers on the national level, given that the majority of FSC certificates were issued in the absence of national standards. According to one business interviewee it remained a weakness of the FSC system that the institution itself was the highest standards-setting authority, regardless of the creation of ASI. Concern was also expressed that the financial relationship between FSC and WWF had the potential to impact on FSC's reputation as an entity independent of specific interests. Buyers' groups created by WWF to encourage companies to use certified FSC timber, for example, meant it had become difficult to disentangle the two institutions, especially when a number of WWF employees also acted as contact persons in various NIs. Two NGO interviewees commented that WWF was now distancing itself from FSC, as it had become uncomfortable both establishing and participating in standards-setting.

Internally, FSC has an array of mechanisms to ensure that the institution itself is accountable at all levels, from the General Assembly to the NIs, and to both members and non-members. Nevertheless, the relationship between certifiers and various parts of the system (clients, FSC itself, national standards bodies), as well as excessively close relations with groups such as WWF, have exposed the system to claims of conflict of interest. FSC has put in considerable effort to ensure accountability across the system, but criticism persists, resulting in a score of **medium**.

Transparency

In 1999 FSC was criticised for its handling of the SmartWood Costa Rica teakwood certification 'as a clear example in which FSC has been unwilling or unable, to comply with basic principles of transparency'.[21] Partly in response, the General Assembly required FSC to identify and recommend best practices for consultation, in collaboration with the membership, and with the approval of the Board, in 2002. Strategic decisions regarding national and regional matters were also to be made 'based on open and transparent consultation with FSC members and the NIs of the region'. Since 2004 certification bodies have been governed by a standard, which details their public reporting obligations. It should be noted that scholars consider that FSC requires more disclosure than other forest certification programmes. NIs, for example, must engage in their activities in a participatory and transparent manner. This is identified as being important for FSC to maintain its credibility as a total system. Meetings and sub-committees are open to all interested parties, members or otherwise. Non-members may attend as observers and may comment on document drafts. Certifiers must also make their generic standards available in national or local languages when undertaking consultations.

NGO interviewees argued that because FSC was open to scrutiny it attracted more attention than other more closed programmes, and was in a sense the victim of its own high standards. Nevertheless, concerns were raised by both by NGOs and business interests about the lack of transparency in the use and application by certification bodies of interim standards on the national level. Business informants and members of the third group of interview subjects were also concerned about the lack of transparency surrounding the financial arrangements underpinning the WWF/World Bank Global Forest Alliance, and the promotion of FSC certification.

Generally speaking, however, FSC has put in place a number of provisions to ensure that the institution operates in a transparent manner. The very high degree of transparency at the international and national levels of the institution itself is partly offset by transparency issues at the individual certificate level, particularly those assessed against generic standards. Taking these issues into consideration, an otherwise very high cross-institutional rating is reduced to **high**.

Decision-making

Democracy

The purpose of FSC's tripartite chamber system is to 'maintain the balance of voting power between the different interests, without having to limit the number of members'.[22] In each chamber the membership is further divided into northern and southern sub-chambers with 50 per cent of the voting power each. All members of FSC are represented equally through a weighted voting system. Individual members, to avoid undue influence, are restricted to 10 per cent of the voting weight in each sub-chamber. Commentators have a range of views on democracy within the FSC system. Beyond promoting environmental values, FSC's criteria have been described as advocating democracy and human rights.[23] Other arguments have been made about the appropriateness of placing small forest owners in the same chamber as industrial forest owners, supply chain companies and consultants.

Interviewees had mixed views regarding the nature of the democratic processes within FSC. NGOs considered the approach whereby motions were discussed and amended in all of the chambers was an innovative, albeit slow, way of dealing with previously conflicting parties. A similar degree of sophistication also exists at the NI level, although some interviewees questioned this. One business interviewee felt that the FSC's procedures went against what they saw as the provisions of 'normal democracy', as expressed in the nation-state model of parliamentary democracy. Views from the third group of informants were polarised between those who expressed their concerns about the lack of democracy for private forest owners, and those who saw FSC's 'Brundtland-type' democratic model as being far more legitimate than those that occurred within the nation-state and similar forest-related governmental processes. Despite these negative comments, FSC's procedures nevertheless appear to be capable of balancing many of the divergent objectives of environmental, social and economic interests, and deliver a sophisticated 'triple bottom line' model of democracy, which also accommodates northern and southern, as well as individual, interests. It should also be noted that there are criticisms amongst commentators concerning the level of democratic input for small forest owners and social interests. Overall, however, democracy in the FSC system is **high**.

Agreement

As a norm, General Assembly decisions are made by consensus, understood as the absence of sustained opposition, but not requiring unanimity. In order for a motion to move forward to a vote on the floor a quorum of 50 per cent + 1 of members present and voting within each of the sectoral sub-chambers is first required, followed by a simple majority approval. Once on the floor, a motion again requires 66.6 per cent approval of all the members present and voting (qualified majority). The Board and national and regional initiatives also make decisions by consensus, and specific arrangements are in place where this cannot be reached. Formal FSC Working Groups are expected to reach decisions by consensus. Where this is not possible, decisions must be made democratically, although it is not specified how. Consensus provisions are not universal, however. The selection of National Advisory Boards, for example, may be through either consensus or other democratic decision-making procedures. FSC has been described as the most advanced of the certification programmes in the use of consensus-based decision-making, but it has also experienced some difficulties. There have been instances in Canada when FSC initially accepted what were interpreted as consensus outcomes, but which had not included certain forestry interests. It has subsequently amended its understanding of consensus and now expects industry acceptance of the standards developed.[24] FSC follows the ISO definition of consensus (see below).

NGO interviewees were supportive of FSC's methods of reaching agreement, identifying the chamber system and use of consensus as revolutionary, although time-consuming and costly. One interviewee from the third group of subjects noted how the strong degree of reluctance amongst participants to vote tended to lead to exhaustive discussions that generally resulted in consensus decisions.

The methods used for reaching agreement within the FSC system are interesting. The ability within FSC's deliberative processes to move between simple majority voting in the chambers and consensus may well be the key to the success of the system's decision-making processes, resulting in a score of **very high**.

Dispute settlement

It has been argued generally that FSC has helped resolve more ingrained societal level disputes over forest management. The need for collaboration regarding the development of solutions to contentious issues within FSC system appears to discipline the parties involved. Academic scholars have questioned FSC's capacity to settle disputes concluding that the system was cumbersome and in need of reform. Neither the formal nor the informal processes within FSC system were considered as being adequate to the task of handling the complex dynamics that characterise certification processes. FSC itself acknowledged that its dispute mechanisms were complex, difficult and unclear. A revision process commenced in 2004 and, as a result of agreement at the General Assembly in 2008 to comprehensively review the old Dispute Resolution Protocol, new procedures were adopted in late 2009. These cover dispute resolution, formal and informal complaints about the FSC certification programme generally, as well as appeals against FSC decisions (with the exception of accreditation decisions, which are handled by ASI). These new procedures, it is claimed, will streamline the process and shorten the time for disputes to work their way through the system, aided by a new Appeals Panel, discrete from the Board (unlike the old Protocol). There is also an online service for tracking the progress of disputes through the FSC system, and the previous distinction between member and non-member appellants has been removed. Elaborate dispute provisions also appear in the standards covering controlled wood. Here wood cannot enter the certified supply chain from forest management areas where civil rights have been violated, or substantial – and unresolved – conflict exists amongst indigenous peoples or civil society groups regarding long-term tenure or use rights.

Subjects were interviewed prior to the new procedures, and views at that stage were either critical or ambivalent. Business interviewees were of the opinion that the existing arrangements were both biased and lacking in objectivity. NGOs were generally more interested in the value of FSC as a tool to solve deeper disputes regarding land use and tenure, rather than its internal mechanisms for resolving disputes. However, there is some anecdotal evidence that previously identified historical tensions over levels of compliance remain unresolved. One NGO interviewee commenting on such tensions within their own country felt there was a backlash coming, observing that previously supportive local campaign groups were now actively complaining about every act of non-compliance with the standard.

Reform of FSC's dispute settlement procedures was clearly necessary, as previous arrangements were not well regarded. On paper the new provisions appear to help communications between FSC and those complaining, and make the process more transparent. The new system is yet to be thoroughly tested, but the substantive changes lift an otherwise low rating to a provisional **medium**.

Implementation

Behaviour change

FSC has been described as having a 'norm-building and behavioural effect in the field of sustainable development'.[25] Its governance arrangements utilise both top-down and bottom-up processes of information sharing, created within networks, which foster organisational learning and mobilise resources amongst stakeholders to further improve organisational structures. It is the very diversity of the actors involved which facilitates these learning processes.[26] The institution has also been portrayed as having had a beneficial influence on the relations between the various stakeholders involved in forest policy discussions, notably in those countries where forest governance is weak. In the case of South Africa, for example, FSC certification has been described as bringing previously excluded stakeholders into the national forest dialogue. Elsewhere, in countries such as Bolivia, it has worked as a complementary agent in encouraging greater compliance with national forestry regulation or, as in the case of Mexico, has influenced the development of national regulation, which reflects FSC standards. The continued support and collaboration of environmental and social interests, the endorsement of some state interests, and its influence on national forest policies are all indicators of FSC's success in this regard.[27]

According to one NGO interviewee, forest owners initially found it hard to accept FSC standards and change their practices. FSC helped contribute to a more diffuse change in the forestry community's understanding of sustainability away from equating it solely with sustained yield, to including environmental and social elements. One interviewee from the third group of subjects was particularly enthusiastic about the internal norm-building impacts of FSC's structures and processes, which they felt had generated a high level of trust in the system amongst members.

Given the level of change that is still required to improve forest management in many countries more generally, it is probably best to err on the side of caution and score FSC's impact on behaviour as **medium**.

Problem-solving

FSC has been characterised as a response to the failure of state interests to deliver on binding regulations, which has provided a problem-solving mechanism more suitable to creating regulation than that afforded by intergovernmental agreements. Its model of private governance, notably its use of consensus-based standards-setting and third-party compliance verification makes it an altogether more effective facilitator of global environmental problem solutions. Not all assessments are as positive, and some doubt has been expressed regarding its success in developing countries. By 2005 FSC certified sources constituted 5 per cent of world market share, and 1.36 per cent of global forest cover. These are disproportionately located in North America and Europe (79.2 per cent of certified area), whilst Africa, Asia, Oceania and Latin America only represent 20.8 per cent, and figures are not greatly different for CoC certificates.[28] With only 10 per cent of all tropical timber being certified, FSC's contribution of certification to sustainability is small, but its contribution to sustainable production should still be viewed as a major achievement.[29]

On an environmental performance level, claims have been made that FSC certification is more rigorous regarding forest management and environmental sustainability than other, more 'industry-friendly' schemes. This relates particularly to the protection of old growth forests and the maintenance of biodiversity, as well as restrictions on the use of chemicals, the size of clearcutting operations and an outright ban on genetically modified organisms. Its standards in relation to the conservation of biological diversity have also been emphasised as being the most stringent of all the certification programmes. But in significant supply regions such as Europe, the contribution of FSC certification generally to ecological sustainability is unclear. In the Nordic countries, FSC certification has been less effective than PEFC in promoting SFM amongst small-scale, as opposed to large-scale forest owners, for example. While it is claimed that European forestry practices are likely to improve subsequent to FSC certification, other research argues that certification does not result in substantial modification of existing management.[30] In the case of Germany, for example, one study revealed that 37 per cent of FSC certified forest operations studied did not have to change their management procedures, which raises questions as to what additional benefit FSC certification is making to European forest management.[31] On social issues, FSC has been described as being more demanding than competitor schemes in terms of protecting workers' and indigenous peoples' rights, and for improving the community benefits derived from forest operations. Positive comments notwithstanding, it should also be noted that NGOs commentators assert that if FSC wishes to retain the confidence of the environmental and social movement into the future, it needs to enforce stricter implementation of its procedures.

Interviewees expressed a number of views concerning the scheme's problem-solving abilities. Both NGOs and business interests had concerns that FSC was in danger of losing its 'edge' over other schemes if it failed to verify whether certification bodies were ensuring that standards were being implemented on the ground. Other interviewees felt that FSC standards focussed too much on forest protection and environmental issues, which had reduced its support in the forestry sector. Both business and NGO interviewees were concerned that FSC had been less successful in improving forest management than they had originally thought it would be. But it was also stressed that FSC had made some contribution to solving global forest problems and had improved forest governance where governments had been unable to do so.

FSC has contributed substantively to the policy debate regarding the improvement of forest management, and has managed to exert some positive influence in the tropics, but as with its impact on behaviour change there are external factors beyond its control – such as the illegal timber trade – that lead to a rating of **medium**.

Durability

FSC has been described as a certification scheme that aims to be responsive to varying socio-economic and ecological circumstances, whilst avoiding the 'excessive' flexibility that business demands might place upon it. Some flexibility is built into the system, but, it has been argued, only to a limited degree; consequently any concessions that economic interests should expect from the system are limited. The difference has been attributed to FSC's reliance on performance-based criteria, in contrast to the systems- or process-based approach of its competitors. There are nevertheless some negative aspects to the FSC system associated with the regional development of management standards, most notably the degree of consistency. In Sweden, for example, an arrangement exists whereby companies that are certified under FSC and PEFC schemes may provide wood to both systems, effectively resulting in equivalency between schemes, despite differences over indigenous peoples' rights and other environmental requirements. In North America, the Canadian Maritimes standard is more rigorous and comprehensive than that developed for the Northeast US. In 2004, clearly responding to concerns regarding inconsistency, FSC released a series of standards with the intention of ensuring that certification, auditing and labelling was conducted in a 'consistent, reliable and credible manner'.[32]

Interviewees had divergent views across all three groups. Some felt that the FSC's chamber system contributed significantly to its capacity to adapt to local conditions, but this also meant the system was unpredictable and slow to adapt. Others blamed the lack of flexibility in standards development, particularly its refusal to explore stepwise, or phased, incremental certification, as one reason why the system had not proliferated as much as it could. Conversely, lack of flexibility was also interpreted as one of the strengths of the system as it guaranteed consistency. Others felt that FSC was already far too flexible, and pointed to the changes made to percentage-based claims as an example. Finally, one interviewee attributed the success of FSC to its longevity, which had given it a comparative market advantage over other schemes.

Broadly speaking, FSC appears to have been able to balance the need for strict and internationally consistent rules with the requirements for local flexibility, although this has not been without problems. It is possible to discern an overall upward trend in the evolution of FSC's structures, processes and products over time, and the organisation has responded proactively to criticisms of inconsistency. In all the main aspects of durability (resilience, flexibility and adaptability) FSC ranks **high**.

Conclusions

Initially developed largely by ENGOs and a few, though important, forest industry players, FSC grew extraordinarily quickly, making certification a particularly appealing and comparatively fast-track avenue for global forest policy change. The particular form of its governance, most notably its market orientation, non-state emphasis and participatory philosophy reflects the emerging discourse of sustainable development.

FSC's NGO origins, and the speed of its development, were to have a number of longer-term consequences, however. Given the historical forest industry suspicion of environmental groups, and the rapidly shifting policy climate at the time, the almost instantaneous development of competitor schemes is not surprising. Secondly, the haste with which FSC's initial certificates were issued, and the credibility problems they generated, was to have a feedback effect on the growth of its rivals. Internally, the growth strategies of the early years also impacted on the organisation's performance, and severely strained its relations with key interests.

Nevertheless, FSC has managed to capture and maintain the support of a wide range of interests. Even though the numerical and geographical representation of these interests may not be equal, the chamber system addresses any potential power imbalances. The institution's decision-making processes also contribute to its success, as they provide multiple entry and exit points for problems to be circulated until solutions are found. The constant interplay between consensus and voting affords participants the opportunity to revisit issues on several occasions. Interestingly, where the system's deliberative processes are weak, the broader opportunities for collaborative dialogue inherent in the system may serve to circumvent blockages.

FSC is not without its failings, however. Several recurring issues are a cause for concern. Firstly, certification bodies have a real potential to act as 'rogue' elements in the system. They are of course meant to be independent, but this creates problems of its own, most notably in relation to accountability. Secondly, the relationship between FSC and some of its core constituents, most notably WWF, has not always been a healthy one. Finally, there is a major unresolved tension in the system between participation, which is largely structural and occurs at the international and national levels, and consultation, which is largely procedural and occurs at the local level.

The recent changes to FSC's governance help clarify and improve the democratic relationship between individual and organisational members. It is also useful to expand the membership system, as this provides more opportunities for the system to demonstrate its organisational responsibility to a broader base than before (as well as increasing revenue). The true value of the new appeals process and the Appeals Panel has yet to be determined.

4
ISO, TC 207 and the 14000 Series

Historical overview

Origins, institutional development and controversies

ISO, derived from the Greek word *isos*, meaning equal, also known as the International Organisation for Standardisation, develops international standards for products, services, processes, materials and systems, as well as for conformity assessment and managerial and organisational practice. International standardisation began in the electrotechnical field with the creation of the IEC in 1906, and was followed by the creation of the International Federation of the National Standardising Associations (ISA) in 1926. ISA's activities ceased in 1942 on account of the Second World War, and in 1946 delegates from 25 countries met in London to create a new body, ISO, which commenced work on 23 February 1947. Its role is to 'promote the development of standardisation and related activities in the world with a view to facilitating international exchange of goods and services and to developing cooperation in the spheres of intellectual scientific, technological and economic activity'.[1] It has developed over 15,000 international standards through a network of 156 national bodies and 580 liaison organisations.

ISO moved into the arena of social and environmental standard-setting relatively recently. The origins of the 14000 Series have been linked to an evolution beyond ISO's traditional product standards to the process standards of the 1980s, exemplified by the ISO 9000 (quality management systems, or QMS) standards. The 9000 Series has been characterised as representing ISO's first efforts to certify management practices, as opposed to compliance against a technical norm.[2] The broad uptake of this standard led many national standards bodies to see the need for a certifiable EMS. The ISO 14000 Series of standards in turn delineated the EMS requirements for firms seeking to become certified.[3] Technical Committee TC 207 was established in 1993 to develop a new series of standards covering environmental management, tools and systems. ISO stresses that

> ISO TC/207 does not set limit levels or performance criteria for operations or products; instead, its activities are based on the philosophy that improving management practices is the best way to improve the environmental performance of organizations and their products.... By providing a framework for improved environmental performance [TC 207's standards] will be contributing to one of the key purposes of environmental management standards:...the goal of sustainable development.[4]

TC 207 held its inaugural plenary session in Toronto in June 1993. The specific importance of TC 207 to social and environmental interests and their lack of involvement in the early stages of standards development led to demands for improved participation. TC 207 first recognised the problem of under-representation in 1998, creating a non-governmental organisation (NGO) Contact Group to ascertain NGO attitudes regarding its work. Based on its findings, an NGO Task Group (TG) was formed in 2001, which produced two reports containing recommendations on how to expand and enhance NGO participation in standards development. In 2003 a higher-level task force was created, the NGO–Chairman's Advisory Group (CAG), consisting of four NGOs and four representatives of the CAG, to review the recommendations of the 2001 TG. Whilst speaking largely to its own internal audience, the NGO–CAG task force nevertheless recognised the importance of NGO participants: 'The quality of any standard is directly related to the depth and breadth of expertise brought to bear in its development. A standard-setting body that fails to engage the necessary expertise is compromising its ability to produce valuable standards.'[5]

In April 2007 the task force published a number of draft recommendations on interest representation within the system.[6] Shortcomings in involving relevant interests at the national level were also identified as part of the task force investigations, and national standards bodies were interviewed to determine their current activities and views. Failure to achieve the best balance of interests, and the lack of

resources, especially in terms of travel and accommodation, were identified as impediments to national-level engagement and consultation.
Inconsistencies in arrangements for interest representation at the
national level from country to country were also identified. Recommendations included addressing the problems associated with balanced
representation (including the role of liaison organisations and national-
level input into ISO's structures), definitions of 'consensus building'
and settlement of disputes over standards development. One of the
most noteworthy proposals was to develop new categories for those
participating in meetings, classified under the new headings of Government, Producer Interest, Service/Professional and General/Public
Interest, and to track the balance of meetings based around these four
categories.

Despite the recognition of these problems, and the suggestions to
address them, a degree of opposition arose in some national standards
bodies. If such changes were to be made, it was argued, they would
have to be applied across all of ISO's TCs. If NGOs were included in
national delegations, they could no longer claim they were representing
the national interest.[7]

These underlying tensions were to come to a head at a meeting of
TC 207 held in Bogotá in June 2008. At an informal event scheduled
before the matter was formally discussed, NGOs gave the impression
that the proposed modifications to the procedures could be seen as a
first step towards the participatory procedures of the ISEAL. A number of member bodies voiced their opposition to such a notion, and
consequently, when the matter came forward for a formal decision, several of them rejected the task force's recommendations. In an attempt
to salvage what amounted to more than ten years' work, it was suggested that a new contact group be established under the control and
direction of the Chair.[8] This was not at all to the liking of NGOs. They
labelled the meeting 'a crushing defeat for consumer and environmental
interests in standardisation' accusing ISO of being dominated by business interests and of 'not fulfilling its strategic policy commitment to
ensure broad and meaningful public-interest participation in advancing its goals of developing standards for a sustainable world'.[9] The task
force, and the process to examine NGO participation in TC 207, were
consequently wound up, although the matter did not end there. A letter
of complaint sent by NGOs to the ISO Secretary General resulted in the
establishment of a group under the auspices of the Technical Management Board (TMB) to look at participation in ISO in a broader context,

taking on board the lessons learned in the development of the ISO 26000 Series. TC 207 will be taking its guidance from the results of that inquiry.[10]

TC 207 and forestry conflicts

Of the TC 207 sub-committees SC1 and SC3 are of most relevance to forest management. ISO 14001, generated by SC1, was at first quickly adopted by the forest industry for certification purposes. By 2003, of the 33,950 ISO 14001 certificates issued across all industrial sectors, 1223 were awarded to the forest product industries, ranking them eleventh out of the sectors seeking certification. At the third plenary meeting of TC 207 in Oslo in 1995 the member bodies of Canada and Australia submitted a new work item proposal to develop an application guide for the forest industry based on ISO 14001. The proposal was intended to create a link between the certification under ISO 14001 and SFM.[11] One impetus for the proposal appears to have been the Canadian forest industries wish to gain international recognition for the CSA standard for SFM, which was being developed at that time.

At the meeting environmental NGOs indicated they were worried that the guide could be misapplied. These concerns can be traced back to the decision that ISO 14001 would not establish absolute requirements for environmental performance, and that companies with different levels of performance could be in compliance. Although ISO 14001 could not be used to communicate environmental performance to the public, there was a fear that EMS certification would nevertheless be used to give the impression that a company had been awarded an environmental label. Consequently 'the issue aroused some controversy' at the Oslo meeting.[12] Greenpeace and other NGOs opposed the proposal on the basis that an EMS was not an adequate measure of sustainability if it lacked performance requirements. They also questioned the adequacy of interest representation in ISO's discussions regarding SFM. ISO members were split between supporters and opponents of the initiative, the latter feeling that the proposal for a sector guide approach would weaken the core EMS documents. By the concluding plenary it was clear that more discussion was needed, and Canada and Australia asked that the new proposal for a formal work item be reclassified as an item for discussion. It was agreed that a broadly based group of interests would discuss ISO's role in SFM over the next year. The group was to include representatives from TC 207, consumer groups, industry and environmental NGOs such as Greenpeace and the World Wildlife Fund (WWF). The Canadian and Australian initiative as it had originally stood was therefore withdrawn.

But the issue was far from resolved. Discussions over the next year took the form of an informal study group established by ISO's New Zealand member body (Standards New Zealand). There was agreement that it would exist outside of TC 207 and meet prior to the 1996 plenary meeting of TC 207. The initiative fell foul of environmental groups from its first meeting. Despite being informal the process was seen as being closely linked to ISO, especially since attendance was limited to ISO members, although the rule was selectively applied. Greenpeace was initially denied access, only to be invited on the day of the meeting. NGOs claimed that their input into the first meeting was systematically ignored – particularly suggestions that sought to expand discussions regarding forest certification beyond the EMS approach. The minutes wrongly claimed that there was consensus that ISO 14001 was a suitable framework for forest certification. NGOs did not participate in the second meeting. The study group submitted a report to the 1996 TC 207 plenary identifying TC 207 as the most appropriate body to address the issue, recommending that a 'bridging document' be developed to provide information as to how ISO 14001 should be applied in the context of forest management. The report was published as technical report (TR) ISO TR 14061 in 1998. Critics interpreted this as a back-door attempt to bring forest certification back under the control of ISO:

> The result would be a range of claims about 'sustainable forest management' that would be very confusing in the market place.... In fact, given the 14001 framework, this bridging document has the potential to allow companies to claim SFM without specifying what performance levels are required. A case could be made that this proposal is indeed a way to bring back the CSA proposal that was withdrawn in 1995.[13]

The forest controversy also played itself out in discussions around environmental labels. NGOs again raised concerns that the possible misuse of ISO 14001 might exempt forest certification from environmental labelling altogether. The reason for such a strategy, it was argued, was on account of timber industry reluctance to embrace environmental improvements that a label such as the FSC would require. ISO 14020, the standard covering the general principles for environmental labels and declarations, had the potential to override all other standards on environmental labelling, currently existing (like FSC) or in the future. This was again interpreted as an attempt to bring forest certification under ISO, since 14020 was designed to be broadly compatible with the forest certification related aspects of the IPF – including the emerging and industry-preferred competitors to FSC. Canada's CSA standard used criteria similar to those of the IPF. Arguments over the guidelines for all types of labelling occurred largely in working group 3, where it appears that NGO initiatives to change the content and direction of ISO 14020 were consistently overruled by narrow margins. This was to lead to an observation that 'WG3 is the perfect example of a working group where consensus has not been achieved.'[14]

Governance within ISO

System participants

ISO membership is broken down into three categories. A *member body* of ISO is a national body considered to most represent standardisation in its country. Only one such member body per country is allowed, and it may participate and exercise voting rights on any policy or technical committee within ISO. Such members are referred to as participating, or P-members. ISO currently has member bodies representing 150 countries. A second category, the *correspondent member*, is an organisation in a country where there are no fully developed standards setting processes. Such a member may not participate actively in policy and technical committees, although they are kept informed, and may use any standards developed in their own country. They may observe, but not vote in the General Assembly. Correspondent members, as well as member bodies, who choose not to participate actively within committees, are referred to as observers, or O-members. A third category, the *subscriber member*, also exists for countries with very small economies, who pay a reduced membership fee, but who are expected to do no more than maintain basic contact with ISO. They may attend General Assemblies as observers. They do not appear in the category of O-members for technical and policy committees.

ISO also has formal relations with a range of international organisations, the intention being to create networks made up of more than national members. These organisations participate in ISO's work in a number of ways. They may influence ISO through its Central Secretariat, or they may be directly involved within an ISO Technical Committee, as advisory experts. They constitute a further sub-category of participant, and are referred to as a *liaison member*, or occasionally as L-members. These include consumer groups as well as industry associations, inter-governmental organisations and UN agencies. Other specific sectors of civil society are also included in this category.

Liaison organisations are themselves broken down into four categories. Category A organisations make an effective contribution to the work of a technical committee or sub-committee. They receive all relevant documents and are invited to meetings. They may also nominate experts to participate in an ISO working group or project team. Category B refers to organisations that wish to be kept informed of the work of a technical committee or sub-committee. They are entitled to receive the reports on the work of any relevant committee. Category C is reserved for the ISO/IEC Joint Technical Committee 1 on Information Technology. Category D is for those organisations that wish to participate in the specific work of a working group or project team. They receive all relevant documents and are invited to working group or project team meetings. Participation of liaison organisations within ISO is convoluted and constrained. ISO's Secretary General, in consultation with the relevant technical committee secretariat, is responsible for approving such liaisons. The status of such liaisons is centrally recorded and reported to the TMB. The procedure for approving such liaison organisations by a technical committee is subject to the consultation and unanimous approval of all P-members in the committee. If not, applications are dealt with on a case-by-case basis.

Individuals cannot be members of ISO, and are reliant on the national member bodies to include them in their participatory structures. Individuals contribute to standards development through selection by ISO members to serve on national delegations to ISO technical committees and they may also provide input in national processes of 'consensus building'.

Institutional arrangements

International level

ISO characterises itself as a voluntary, market-driven, consensus-based, worldwide system, and, curiously, as a NGO. Various institutional entities exist within ISO (see Figure 4.1 below). A General Assembly consisting of ISO members meets annually to make decisions of a broad strategic nature. The General Assembly consists of the officers of the organisation (President, two Vice-presidents, a Treasurer and a Secretary-General) and no more than three delegates per

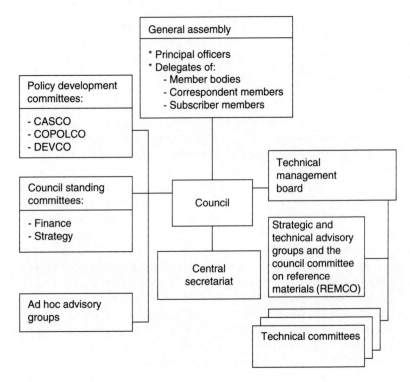

Figure 4.1 Institutional structure of ISO.
Source: ISO, 2008.[16]

member body. Delegations may also contain additional observers, and correspondent and subscriber members may also attend as observers. Each member body has only one vote. Resolutions are adopted in the General Assembly itself or via letter ballot by majority vote. ISO's general orientation is currently driven by its *Strategic Plan 2005-2010*, which was approved by the General Assembly in 2004. The proposals put to the members of the General Assembly are developed in advance by the ISO Council, which ISO itself describes as a business-style board of directors, consisting of representatives from the membership as a whole. Three policy development committees – conformity assessment (CASCO), consumer matters (COPOLCO) and developing country matters (DEVCO) – and two standing committees – finance and strategy – report directly to the Council. The Council meets twice a year, and the positions within it are rotated. It is chaired by a president, who holds the position for two years and who is 'a prominent figure in standardisation or business'.[15]

ISO's Secretary-General reports to the Council and operates out of a Central Secretariat in Geneva of approximately 160 staff.

The technical work associated with standards development is overseen by the TMB, which is responsible for establishing technical advisory groups (TAGs) and technical committees (TCs). The TCs are created to address the needs of particular industries or general areas, to assist in the development of relevant standards. The TMB consists of a chair and 12 member bodies, either appointed or elected by the Council. TCs may also set up further advisory groups, study groups, ad hoc groups and editing committees to support their activities. Decisions concerning a committee's overall policy and strategy are made at the technical and sub-committee levels. These bodies are managed by a committee secretariat, provided by one of the ISO members. The working group level is reserved for researching and drafting, and managed by a convener. Committee secretariats are required to be neutral, and must consist of a P-member of the committee. Committee secretaries work with the chair in managing the committee's work programme.

National level

Member bodies are permitted to use ISO standards as a basis for their own national standards. They attempt to follow the development of international standards domestically through the establishment of what are generally referred to as 'mirror committees', although the exact nature of these committees varies by country. A national member body may create a TAG to develop national-level positions for consideration at the international level, and may send three member body delegates and other observers to attend international technical committee meetings, as either a participating (P) or observer (O) member. NGOs do in some instances participate both in national standards bodies and their TAGs. They may also be present within national member delegations to ISO meetings.

Standards development

ISO 14000 does not exist as an individual standard; rather it is to be understood as a series of environmental management standards describing the basic elements of an EMS. The standards have been and are being developed under ISO Technical Committee 207. Only one, 14001, has been developed for the purposes of third-party certification. The development of international standards is a decentralised

Table 4.1   Stages in the development of ISO International Standards

Stage name	Product name	Acronym
Preliminary stage	Preliminary work item (project)	PWI
Proposal stage	New proposal for a work item	NP
Preparatory stage	Working draft(s)	WD
Committee stage	Committee draft(s)	CD
Enquiry stage	Draft International Standard	DIS
Approval stage	Final draft International Standard	FDIS
Publication stage	International Standard	IS

Source: ISO, 2005, p. 11.

process, undertaken through technical committees at both national and international levels and involving stakeholders across the ISO network. Standards development follows several recognised stages (see Table 4.1 above).

TC 207

Located in Canada, TC 207 is one of the largest technical committees within ISO, producing (and continuing to oversee) the ISO 14000 Series. TC 207's sub-committees and associated working groups have been responsible for managing the generation and revision of a number of standards under the 14000 Series (see Figure 4.2 below).

ISO 14001 is the only standard designed within the Series for registration (another term for certification) by a third party. Alternatively, a company can use the specification for internal guidance only, or for self-declaration purposes, and may choose not to seek third-party verification. These two standards were first released in 1996. Neither internal nor external audits require the evaluation of a company's environmental performance. ISO 14001 has undergone revision and now exists in a version published in 2004 (ISO 14001:2004).

ISO 14001 has five main sets of requirements: the adoption of an environmental policy supported by top management (commitment and policy); the formulation of a plan to fulfil the policy (planning); allocation of responsibilities and associated training and awareness raising (implementation); development of a system for checking, correcting, monitoring, reporting and preventing environmental impacts (measurement and evaluation); and the establishment of management review and associated continuous improvement processes (review and improvement – see Figure 4.3).

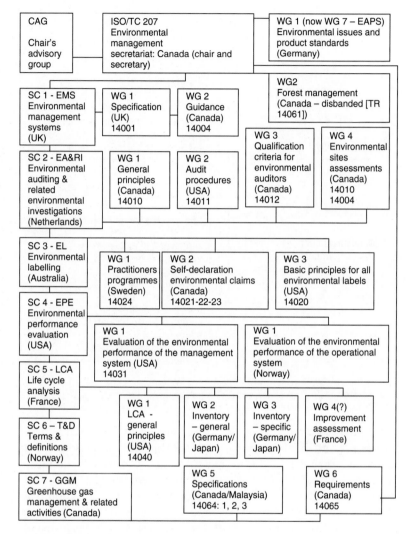

Figure 4.2 Institutional structure of TC 207.
Source: Hauselmann, 1997, p. 7; ISO, 2008 (adapted).

Institutional typology

ISO as an institution has been portrayed as constituting an 'indirect monopolist global governance arrangement'.[17] Other commentators consider ISO variously as a non-governmental, mixed private, or hybrid

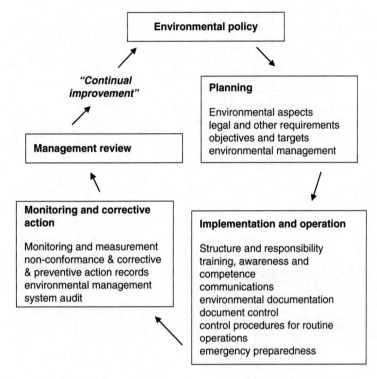

Figure 4.3   The Environmental Management System model for ISO 14001.
Source: Edwards et al., p. 15; Elliott, 2000, p. 15 (following ISO 14001).

public-private regime.[18] Under such an arrangement, government influence is present, but it is not direct, nor all-controlling. Some commentators also refer to the development of ISO 14001 as occurring either in the shadow of public law or, more generally, in the shadow of hierarchy, referring to the role of government in certification.[19] In view of these observations, it has been identified as a governance system that sits closer to the non-state, rather than the state-centric end of the sovereignty continuum, but only slightly. National standards bodies have a degree of interest in and influence over ISO's standards given the often-close linkages of the latter to state regulations and international trade rules. Some member bodies may be state institutions, or if private, receive funding from the state, and it is at the country, not organisational level, whence ISO derives much of its authority. The strong role of industry inclines the institution towards the non-state end of the axis as well, but this is itself mitigated by the subservient role played by other non-state interests, such as NGOs, and the close alignment of business and 'quasi-state' standards bodies. Consequently, it is located on the non-state end of the authority axis, but with a rating of **low.**

Although ISO is located towards the deliberative end of the democracy axis, it is still effectively an institution that functions along aggregative democratic lines. There is undeniably an emphasis placed by the institution on the use of consensus in its public literature, and the qualified majority voting associated with standards approval, which requires 75 per cent agreement, and this should imply a high degree of deliberation within its decision-making bodies. This is offset, however, by the use of the aggregative democratic method of voting by simple majority

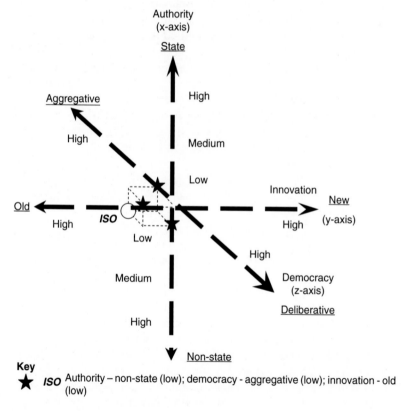

Key

★ *ISO* Authority – non-state (low); democracy - aggregative (low); innovation - old (low)

Figure 4.4 Institutional classification of ISO.

throughout most other parts of the institution, and the discretion given to technical committees as to whether they use consensus or majority voting. Added to this is the inconsistent use of voting and consensus within TC 207, the main sub-institution studied here, in its committees and working groups. ISO consequently sits along the aggregative end of the democracy axis, but acknowledging the use of consensus in some associated entities, it is rated **low.**

Given its monopolist, rather than collaborative multi-stakeholder arrangements, it is difficult to view it as representing a particularly 'new' model of governance. Some commentators, however, do see the institution as being an example of 'mixed' mode governance, referring to its private–public nature. However, to this should be added the caveat that the organisation's membership is fairly exclusively defined – and confined – by traditional territorial boundaries. This membership effectively interacts in a way not unlike member states of the UN, using a one-country-one-vote method of reaching agreement. The placement of industrial interests in positions of power within its institutional structure, and the commensurate lack of a role for NGOs, also leads to the conclusion that ISO is not especially innovative in its institutional design. Recent developments, most notably the attention paid to NGO representation on a formal level (if not yet fully realised) indicates some willingness to change. Consequently it has been located on the 'old' end of the innovation continuum, but is rated **low** (see Figure 4.4 above).

Governance quality of ISO

Commentary

ISO, TC 207 and the 14000 Series in particular received 30 points out of a maximum total of 55. Four indicators achieved low ratings (equality, transparency, dispute settlement and problem-solving), and one a high rating (durability). On the criterion level the conventional pass/fail target value of 50 per cent for interest representation (53 per cent), decision-making (53 per cent) and implementation (60 per cent) was exceeded; organisational responsibility reached, but did not exceed, the target value. At the principle level of meaningful participation, the combined values of the two generated a result of 52 per cent, exceeding the target value of 50 per cent; in the case of productive deliberation the target value was also exceeded with a result of 57 per cent. A total score of 55 per cent was achieved overall. See Table 4.2 below.

Table 4.2 Evaluative matrix of ISO TC 207 governance quality

Principle	1. Meaningful Participation					
Criterion	1. Interest representation Highest possible score: 15 Lowest possible score: 3 Actual score: 8			2. Organisational responsibility Highest possible score:10 Lowest possible score: 2 Actual score: 5		Sub-total (out of 25): 13
Indicator	Inclusiveness	Equality	Resources	Accountability	Transparency	
Very high						
High						
Medium	3		3	3		
Low		2			2	
Very low						

Principle	2. Productive deliberation					
Criterion	3. Decision-making Highest possible score: 15 Lowest possible score: 3 Actual score: 8			4. Implementation Highest possible score: 15 Lowest possible score: 3 Actual score: 9			Sub-total (out of 30): 17
Indicator	Democracy	Agreement	Dispute settlement	Behaviour change	Problem solving	Durability	
Very high							
High						4	
Medium	3	3		3			
Low			2		2		
Very low							
Total (out of 55)							30

Evaluation

Interest representation

Inclusiveness

Under ISO rules member bodies must take into account the full range of interests relevant to the development of a given standard. This is expected to entail the arrangement of national consultations and the preparation of national positions representing a balance of interests. Some national rules provide for consultation and participation of NGOs. Other member bodies do not formally exclude NGOs, but neither are they obligated to consult with them. The national member bodies within the development of ISO 14001, for example, have been criticised for their low NGO participation rates. The conversely high presence of industry groups is deemed inappropriate by some commentators. The historical extent to which NGO interests have been included in TC 207 has been identified as a pressing problem. Representation in plenary meetings between 1997 and 2003 remained relatively unchanged, and low, at only 3 per cent of delegates, in comparison to the top four attendees: industry 32 per cent, standards organisations 21 per cent, consultants 18 per cent and governments 10 per cent.[20]

Developing countries' participation in ISO is also a historical problem. The need to enhance the representation of developing countries' interests within ISO has been officially recognised since 1961, when the organisation created a policy committee, Developing Countries Committee (DEVCO), to look into the needs of developing countries in relation to standard-setting and related activities. In 2002 ISO created the Developing Countries Task Force (DCTF) to identify specific responses to the issue of under-representation. ISO 14001 in particular has been described by scholars as 'a self-regulatory international environmental management standard created by and for industry, but without the full participation of weaker stakeholders in the international community'.[21] Some further corroborative evidence is found in a survey of 55 certified companies in the US conducted in 1998, which showed that 67 per cent of ISO 14001 certified companies had not involved community stakeholders in developing or implementing their EMSs.[22]

The views provided by interviewees regarding ISO's inclusiveness were mixed. Business interests appeared generally happy with their involvement, partly because of their direct experience of implementing ISO 14001, and on account of their role in shaping the course of the Series' application nationally. NGOs believed that the extent of their sector's involvement within ISO was marginal, which was the reason for their focus on trying to change the governance of TC 207 and across ISO more generally. They were highly critical of the move by national standards bodies to reject the NGO–CAG task force proposals, arguing there was no point in pursuing alternative measures. NGOs were now waiting for the results of an ISO TMB examination into ISO 26000's more innovative arrangements for stakeholder involvement. These would effectively supersede TC 207's failed procedures. Other interviewees, notably those from the third group, defended ISO's degree of inclusiveness. TC 207 was trying to engage with NGOs more effectively and had created processes to do so, but there was resistance from some committee and sub-committee chairs. Some acknowledgement was given to efforts being made to include developing country stakeholders more effectively through the 'twinning' process of having co-chairs from developing countries, and by giving them more responsibilities within the ISO system, but it was also acknowledged that sub-committees tended to represent more important countries. On a national level it was claimed that there were examples of countries that had a very open systems of participation. It was stressed that ISO was beginning to see the need for better engagement with all interests. However, different countries and processes were at different stages of development. One NGO also commented positively on their level of involvement in standards development at the national level.

ISO is characterised by a considerable amount of variation in the extent to which different interests are included in its representative structures. Members, who essentially represent national interests, are themselves divided into three different categories, only one of which is entitled to vote in the General Assembly and in the technical committees and sub-committees associated with standards development at the international level. Civil society interests, particularly NGOs, are even less included internationally. The degree to which different interests are represented at the national level also varies from country to country. This affects the country positions adopted and presented at the international level as articulating the national consensus. In terms of TC 207, earlier positive efforts to include NGOs more effectively have now ended in stalemate. Another civil society inclusiveness problem arises in relation to low levels of consultation associated with company implementation of ISO 14001. The positive developments within the area of social responsibility now appear to have outpaced TC 207, and the move to reject the NGO-CAG task force report can only be characterised as reactionary. ISO's inclusiveness is **medium.**

Table 4.3   ISO member bodies and technical committee participation by ISO geographical region

Region	ISO member bodies (%)	Average TC participation (%)	TC Secretariats (%) (NB: 4% vacant)	TC Chairs (%) (NB: 18% vacant)
Western Europe	12	48	58	47
North America	2	7	19	22
Asia	25	22	10	8
Central and Eastern Europe	15	13	3	2
Africa	30	4	3	2
Oceania	2	3	2	1
Central and South America	14	3	1	0

Source: Morikawa and Morrison, pp. 9–13 (as of October 2004).

Equality

The make-up of ISO member bodies by region and their participation in standards development within ISO's technical committees is uneven and geographical representation continues to be overwhelmingly from Western Europe and North America (see Table 4.3 above).

Developed and developing country delegates have also commented that the needs of small and medium enterprises (SMEs) were almost completely overlooked during the Series' early stages; this required a great deal of retrospective changes as the standard was developed. Commentators also observe that standards organisations and consultants, 'who are not the primary users of the standards, but the ones who thrive on standards-related business... have more presence in ISO 14000 standards development than industry, the audience for whom the standards are presumably intended'.[23]

The increasing professional and technocratic dominance across a number of TC 207's sub-committees generated concerns across interview groups; it was suggested that the philosophy of small government had resulted in a shift of governmental employees into the private domain, resulting in a numerical bias of consultants. All three groups of interviewees acknowledged the role played by NGOs in forcing ISO to address the imbalance in interest representation. One NGO interviewee commented that some progress could be seen in the Working Group on Social Responsibility but overall business interests were still in the majority. Business also allied itself with member bodies, overwhelming civil society viewpoints; both groups had combined to reject the NGO–CAG task force proposals, for example. International NGOs were also only allowed to participate actively at the lowest, working group, level. Higher-level committees could simply overturn any decisions made at that level. Another commented that in their experience ISO guidance on balanced interest representation was not followed at the national level. Their organisation did not want to become 'alibi participants' and it chose not to participate in national standards development as a result.

Interviewees from business and the third group of informants recognised that national standards bodies and business had the potential to dominate under existing arrangements. Large companies had much more to gain from certification under 14001 than SMEs, which might explain the lack of SME involvement in the system. It was also acknowledged that there was no specific guidance from the Central Secretariat to its ISO member bodies on how participation should be made effective, despite the requirement in the ISO Statutes for balanced stakeholder consultation nationally. ISO had accommodated non-state interests at the international level, but had put pressure on its current country-representative structure. There was a fear amongst national member bodies that introducing new groups would upset the existing balance, especially if the one-country-one-vote rule were abandoned.

Business, standards bodies and certifiers in combination have a much greater degree of influence than all those interests that make up civil society. They are well represented within member bodies domestically and they are also present on national delegations, whilst the number of NGOs is considerably lower, and in a minority. Only members may chair technical sub-committees, and here there is a further problem within ISO's membership associated with western European and North American member domination. ISO's current participatory structures by their very nature favour certain stakeholders over others. The President of the ISO Council is expected to be a leading business or standards figure, and the structure of Council itself follows a business model. Economic interests sit within ISO's executive structures. International NGOs, the main civil society protagonists, cannot occupy many of the positions of power that business and standards bodies can, since they have no membership rights. NGOs can be on national delegations, but they are in a minority. Equality in ISO is **low**.

Resources

According to interviewees, ISO's national members pay subscription fees, in proportion to a country's trade figures and gross national income. The fees are used by ISO to meet the operational costs of running the organisation's Central Secretariat. In 2005 the Central Secretariat had a budget of about 40 million Swiss francs per annum (approximately $38 million). Over half of ISO's national standards bodies are government-funded departments; others are funded by sub-scriptions from industry and other interests. Membership fees at that time generated approximately two thirds of income, while commercial revenue, from the sale of standards, copyright royalties, and so forth, provided the rest. More detailed financial reporting was not made available.[24] The other costs of the system are borne by member organisations through their active management of and participation in standard-setting. Some effort has been made in ensuring developing countries are resourced to participate in ISO activities. Every three years, DEVCO carries out programmes for developing countries (DEVPRO), aimed at increasing informational capacity through training and the publication of training manuals, and providing travel assistance to technical meetings of ISO. Participants in ISO's own executive organs, such as the Council, are expected to cover their own travel and subsistence costs.[25]

The costs associated with participation in ISO were a major preoccupation across all groups of interviewees. In the international standards setting arena it was acknowledged that participants attended at their own expense, and they did not attend if they were unable to do so. Travel and accommodation costs for such groups to participate in ISO meetings were described as extremely high, and NGOs contended that costs associated with international meetings were too much even for industry to pay. The ISO Central Secretariat spent approximately $2 million each year sponsoring developing country participation and had funded a representative from Ghana to TC 207. Mention was made of one other fund established by ISO to assist consumer community groups to help defray their costs to participate. The Dutch government also provided funds for developing country participation. Canada sponsored and paid for NGO and civil society delegates to represent the country in international standards. One NGO commented that much of the power in ISO resided at the national level, but it was here that NGOs lacked the economic capacity to participate.

Access to the resources necessary for meaningful participation is uneven amongst ISO's participants. On the membership level, developing countries struggle to participate effectively in ISO's standards setting processes on account of insufficient funds. However, ISO as an institution has put some mechanisms in place to assist developing country members, particularly through DEVCO and the twinning concept. For non-member groups, such as liaison organisations, there are generally no funds available on an international level, although there are exceptions. The high costs associated with international standards development challenges even the ability of some businesses to participate, particularly SMEs.

There are also problems associated with access to resources on a national level. The fact that some of ISO's member bodies finance standards development on a user pays basis may lead to a participation gap of under-resourced groups such as NGOs, which is created by an inability to cover participation costs. Given these mixed results, the provision of resources to participate is **medium.**

Organisational responsibility

Accountability

In some ways, the TMB, rather than the General Assembly, Council or Board, has the highest degree of responsibility for the governance of ISO, as it appoints chairs, monitors work progress and oversees the implementation of the organisation's rules associated with the development of standards, known as the *Directives*. TC 207 in particular has acknowledged that the move to what are effectively public policy standards has required greater levels of accountability than ISO's more traditional, technical, standards. Nevertheless, an accountability problem regarding the exact status and nature of delegates participating in the development of the ISO 14000 Series has been identified, which has implications beyond TC 207. Country-member delegates (who may actively participate in standards development) are referred to by their country of origin not their organisational affiliation. For example, a delegate from France working for an electronics company is not listed according to their sector, but simply as a French delegate.[26] This undermines a fundamental principle of public or private governance, namely the ability to hold an individual accountable for their actions through a clear allocation of responsibilities and a clearly defined role. This also affects the institution's transparency (see below). It has been contended that a lack of specific directions from ISO means that national member bodies are accountable only to themselves in determining stakeholder participation at the national level.[27]

Comments regarding the accountability of ISO were confined largely to NGOs and the third group of interviewees. Both expressed concerns about the roles played by both consultants and certifiers in the ISO system. In the case of consultants, it was not clear to whom they were ultimately responsible. This would not be particularly important if they were solely engaged for their technical expertise, but they tended to be hired because they knew how to manipulate the system for the benefit of their clients. Certifiers were effectively 'pushers' in the system. As with the quality-related ISO 9000 Series, an industry had grown up around the ISO 14000 Series. It was claimed that certifiers had 'flogged it to death' on the basis that they would certify companies who could then differentiate – and promote – themselves in the marketplace as a 'green' company, when this was not the case.

ISO itself was seen to be relatively accountable by one NGO interviewee. The argument was presented that standards bodies should be seen as the people most interested in ensuring the system was objective, democratic and transparent. While the system might not be particularly environmentally minded or participatory, the institution as a whole certainly had an interest in due process.

As an institution, ISO displays a degree of accountability towards its constituents, although the variation in national member bodies' institutional arrangements regarding participation affects their accountability to national stakeholders. Privately constituted standards bodies may consider paying clients more important than community interests, whilst state-managed institutions may deem the mandate of an election as a sufficient level of accountability. A degree of concern was expressed by a number of stakeholders on the role played by certifiers. This relates to their significance in overseeing the implementation of ISO's standards and the economic nature of their relationship with companies seeking certification. It is encouraging to see ISO placing some emphasis on developing social responsibility initiatives to render ISO more accountable to non-members, although TC 207 has gone backwards in this regard. The need for standards bodies to raise funds through the sale of standards or via sponsorship from government or industry, and the financing of NGO participation by industry and government, has the potential to lead to conflicts of interest. The rating is **medium.**

Transparency

ISO aligns itself to the WTO and that organisation's requirement for transparency and openness to be observed by international standardisation bodies. According to ISO its international standards embody 'the essential principles of global openness and transparency...safeguarded through its development in an ISO Technical Committee...representative of all parties, supported by a public comment phase'.[28] The literature, on the other hand, is critical of ISO, and identifies a lack of transparency in public reporting in two regards. Internally, the organisation is criticised for the very limited nature of publicly available information regarding the scope of representation in standards development. Because there is no systematic approach to monitoring participation it is not possible to determine whether input from a full range of interests has been achieved. This is reinforced by a US study, which found 67 per cent of ISO 14001 certified companies did not publish an annual environmental report of their activities.[29] Further, a company is not required, as with the EMAS, to provide a validated statement describing the company's environmental impacts. [30] It should be noted, however, that ISO 14001 was revised in 2004 with the intention of providing clarification on some of the matters that were unclear in the first iteration. ISO 14001:2004 places an emphasis on the need for a company to identify its significant environmental aspects. An EMS is also required to be legally compliant. A further transparency initiative requires a company to provide documentary evidence of its EMS.

NGOs were the only group of interviewees to comment on the transparency of the ISO system, claiming that it was not developed in an open process at all. In ISO documents were made available to people in the various committees, but during the public enquiry phase the documents had to be purchased by other interested parties if they wanted to make a comment; in many cases the public did not even know that the documents were available. They contrasted this with the Global Reporting Initiative (GRI), where all discursive transactions were undertaken through the Internet, and everybody was afforded the opportunity to comment upon them. Within the GRI system there were clear instructions on how to develop indicators and how to communicate them to the public. Another NGO interviewee provided a telling example of lack of transparency. This anecdote is worth reproducing in full:

When the first ISO 14001 standard was developed, I went to a bridging committee meeting, which was my first direct experience of the standard. There was one representative in particular there who opposed me constantly. I used to address him as 'the American gentleman' until he came and told me he was Canadian. I kept wondering what country he was representing; obviously not Canada, as

this was a purely European event, and only representatives nominated by European Community standard bodies were admitted. It
turned out that the Belgian standards body had nominated him, and
that he was from Exxon. I got into the habit of sitting next to an
Italian and we eventually became friendly. He was from Exxon as
well. I didn't think that was how experts got nominated. Years later
I was talking to the committee secretary and I asked him whether
he remembered the two Exxon people on the committee. 'Two?' he
replied, 'There were four'!

It is difficult to consider ISO on either a national or international level as
being particularly transparent. The development of standards is largely
an 'insider' process, and documents are generally not available to the
public until the enquiry phase. ISO 14031 is to be negatively contrasted
with the GRI in this regard, for example. Given the number of standards
being developed by ISO, and the decentralised nature of their development, the accessibility of information is questionable. Another problem
relates to whose interests given stakeholders are representing when they
participate. The constituencies of the various civil society groups are
quite obvious, but this is much less so in the case of business interests
participating in national delegations or at the technical committee level
and below. Consequently, the transparency of ISO is **low.**

Decision-making

Democracy

It is important to understand the difference in status within ISO between
members and other stakeholders, as this results effectively in two classes
of participant in ISO: member bodies with voting rights; and those without voting rights, which participate in ISO's processes more generally.
An O-member, for example, does little more than receive information
and watch from the outside. Liaison organisations may participate in
discussions and may receive all information, but they are not entitled
to vote in the balloting that is associated with advancing standards
through ISO's drafting stages. TC 207 has recognised that external and
internal stakeholders are likely to have differing perceptions regarding
the effectiveness of the 14000 Series, but has stated that its concerns
lie with its internal stakeholders. In this context effectiveness is understood in a democratic context as overcoming 'obstacles that reduce
a stakeholder's opportunity to influence the decision-making process
relative to other stakeholders'.[31]

Judging by the extent of criticism, ISO's democratic processes appear to have impacted most negatively upon NGO interests, one of whom commented that NGOs did not have any rights within ISO, merely the right to participate. Within TC 207 the exercise of choice was usually by majority vote and minority interests were unable to outvote industrial interests. One NGO interviewee associated with the development of ISO 14001 could recall only one instance when their viewpoint carried the day, and this was only because voting interests supportive of the NGO position had sufficient numbers at the time of voting. Business interests were generally positive. One viewed the current ISO rules of procedure as sufficient. If the rules were changed it would give NGOs undue power, undermining democracy, as it was traditionally understood. Another interpreted ISO as having an electoral system dependent on the nation-state. Only technical people from the various ISO countries could vote, and as they were linked to government, they effectively represented the state. One interviewee from the third group of subjects showed some sympathy for the NGO viewpoint. ISO's reluctance to change was indeed creating problems for the civil society interests that had appeared over the past 20 years. These interests operated at an international level and had a supranational focus, giving them a global understanding and focus on particular issues, but they were not able to bring their experience to the fore because they could not vote.

For those members of ISO who can vote it is fair to say that decision-making is highly democratic. Standards are developed through a multi-stage process, and reflected at the national level through mirror committees. However, voting is a small part of the deliberative processes associated with the creation and publication of standards within ISO. Looking at TC 207 in particular, there appears to be a considerable degree of inconsistency exercised by those in positions of authority in the various sub-committees and working groups over which democratic procedures should occur. The tension here is between the high levels of democracy in the balloting standards and for voting members, versus the dubious nature of the exercise of democracy within some of the sub-committees and working groups in TC 207. At best, the democracy of decision-making in ISO should be considered **medium.**

Agreement

Resolutions at General Assemblies and Council meetings, as well as the initial determination by member bodies to establish new technical committees, are adopted by a majority vote. Technical committees, unless otherwise determined by the TMB function under consensus agreement of the member bodies actively participating, but there are no specific directions as to how such national consensus should be attained. Confusingly, given their status, liaison organisations are also expected to agree to decisions made. Although they cannot vote, technical committees and sub-committees are expected to seek the full and formal support of category A organisations for each international standard in which the organisation participates; but there is no clarity over what full and formal backing means. Under ISO rules member bodies are expected to present a national consensus position to the appropriate technical committee. The high level of business interest participation in decision-making has been challenged, since it raises questions regarding the real nature of consensus within national standards bodies. The extent to which standards development is driven by consensus – as opposed to voting – is unclear, as the procedures during the committee stage of standard drafting indicate. Under its agreement with WTO, ISO must implement principles of 'good governance' including consensus. The role of consensus is stressed in ISO documentation. At present ISO defines consensus as:

> General agreement, characterised by the absence of sustained opposition to substantial issues by any important part of the concerned interests and by a process that involves seeking to take into account the views of all parties concerned and to reconcile any conflicting arguments. Note: consensus need not imply unanimity.[32]

In the development of the 14000 Series gaining consensus in this manner proved elusive. There are several examples where consensus was replaced by a majority vote and issues affecting the broader strategic direction of environmental standards were defeated by narrow margins. This led one commentator to note that the very fact that voting systems were employed demonstrated that consensus was not in operation. The closeness of some votes resulted in a situation 'far from the "absence of sustained opposition" required under ISO rules to achieve consensus'.[33]

Both business and NGO interviewees were critical of ISO's existing arrangements, particularly the inconsistency in the use of consensus within the institution, and the implications for democracy. One business interviewee criticised ISO's decision-making processes for being overly bureaucratic, whilst not actually delivering particularly effective decision-making processes. The third group of interview subjects provided conflicting analyses of ISO's processes for reaching agreement. It tried to follow consensus, but at the end of the day resolutions were accepted on a majority basis. The confusion between whether the decisions were taken by consensus or by voting was clarified by one interviewee as follows: 'The standards setting as they say is by consensus, and there is a definition of consensus, but the practical answer is we go through several rounds of voting.' Decision-making at the national level was portrayed as a process whereby conflicting parties could voice their opposition, but majority decision-making was fairly common.

ISO contains a wide array of mechanisms for reaching agreement, from simple majority to qualified majority, and from consensus to compromise. The exercise of consensus, however, despite WTO requirements, appears to be more honoured in the breach than the observance and there is also some confusion as to when and where standards development should be driven by consensus or voting. The rating is **medium.**

Dispute settlement

In terms of dispute settlement, ISO has three points of entry for those entitled to complain. In technical work, only P- and O-members of a national body may appeal, as indicated above, in a system whereby the complainant goes to increasingly higher bodies until the dispute is settled (i.e. from sub-committee to technical committee), thence to the TMB, and ultimately the Council. The decision of the Council is final. Complaints regarding standard-setting largely relate to procedural matters only, unless matters of principle or the reputation of ISO is challenged. These disputes may be taken by a P-member to the committee chair for resolution, and if unsuccessful there, to the TMB. Liaison organisations appear to have little ability to have grievances addressed within the ISO system. ISO does not have a system to address complaints or disputes regarding the conformity of assessment by third parties or organisations using standards. Here the authority rests with the certification body awarding the certificate. Complaints at this level are expected to be directed to the body that accredited the certifier.

Interviewees from the third group of key informants were the most positive regarding ISO's approach to dispute settlement processes, which was characterised as using both formal and informal methods. Within TC 207 informal dispute settlement measures were more often used. This appeared to work most effectively when the dispute in question related to operational matters, rather than substantive issues, which could at times be heated. Disputes were eventually sorted out through compromise arrangements. Alternatively, matters were worked out through repeated meetings; rather than seeking a two-thirds majority a committee chair might on occasions allow deliberations to continue until almost every national delegation agreed. The informal approach used in TC 207 elicited mixed responses from NGO subjects. Only one was able to recall an instance where conflict had resulted in a beneficial change as the result of extensive debate and discussion. Another interviewee commented that consultants acting on behalf of industry or government often manipulated the nature of disputes, using the pretext that a contested political issue was a technical matter and not 'feasible', thus falling back on a procedural nicety to gag further debate.

Disputes within ISO appear to be managed by passing them higher up the command chain if they cannot be settled at source. Most disputes in TC 207 appear to be settled largely informally, leaving tensions unresolved. Given that there are dispute settlement measures in place, but that informal approaches appear to be the preferred model, the score is **low**.

Implementation

Behaviour change

Discussions in the scholarly literature are generally confined to the impacts of ISO 14001 on changing corporate behaviour. ISO 14001 was sponsored by an organisation with a heavy input from multinational corporations and environmental NGOs are justified in their suspicion of its self-regulatory approach. The conditions under which companies are eligible to join the scheme are extremely broad: companies with a poor compliance record can join up so long as they can afford to establish and maintain a certifiable EMS. Firms can join ISO 14001 opportunistically without following its mandate, since they lack the incentive of having invested in any assets specific to the scheme itself.[34] Although ISO 14001 does require annual rectification audits, it does not appear to have any methods to sanction members who fail to comply with standards. Firms are not required to demonstrate improvements in regulatory compliance to stay accredited, and there is no guarantee companies will make use of the materials collected in the EMS process to actually improve performance. ISO 14001 merely seeks their commitment to do so, and considers the evidence that a company has established and maintained its EMS as sufficient.[35]

Observations regarding ISO 14001's inability to change behaviour were confined almost exclusively to NGO interviewees. Because it did not differentiate between environmental pioneer and the heavy polluter, both could be certified under the same standard; heavy pol-luters could use it for promotional purposes to make the world believe a company had a high environmental performance when it did not. This was a problem when the system was exploited as an instrument of qual-ity assurance. A company might simply choose some easy action, such as changing room temperature by one or two degrees across the company, and although this could only be done once or twice a company could nevertheless demonstrate continuous improvement. In this way it could avoid the need to undertake stronger environmental protection mea-sures. ISO 14031 (environmental performance evaluation) compared particularly poorly with other standards bodies in this regard. Many companies preferred the GRI guidelines, rather than the ISO standard, which said 'everything about the inherent quality' of the ISO standard. The final NGO criticism was that the decision to have a thorough sys-tem was at the discretion of the company, and this decision reflected on the company itself, not the standard. One business interviewee echoed this observation.

Whether companies change their behaviour towards the environ-ment as a consequence of being certified under ISO 14001 is disputed. There is even some evidence to suggest that companies may seek ISO certification to avoid the more onerous constraints of other systems. Behavioural change can certainly be associated with having an EMS in place, but this does not have to be an ISO one specifically. The degree of behaviour change also seems to depend on the company, not the sys-tem. The decision to orientate the ISO 14000 Series of standards around process outcomes, rather than performance has attracted both leaders and laggards. The uptake of ISO 14031, relating specifically to environ-mental performance, has been low. This does not speak well for the performance-enhancing benefits of the system as a whole. It is difficult to determine whether the score should be low or medium. The deci-sion here is to err in favour of ISO and consider behaviour change as **medium.**

Problem-solving

ISO 14001's lack of mandatory external performance-based reporting has been cited by critics as proof that certification 'will do little to avert the historical trend of global environmental damage closely associ-ated with transnational industrial activity'.[36] Companies certified under ISO 14001 do not implement standards in the same way, and there are significant differences as to how they manage their environmen-tal operations. There is some evidence to support the view that firms that certified with ISO 14001 between 1995and 2001 tended to have a lower environmental performance than their peers. Contrary to the expectations of some of the institution's creators, firms with a poor environmental performance were more likely to certify with ISO 14001 than other, more stringent schemes.[37]

A decision was made to remove compliance and performance-auditing measures from the ISO 14000 Series in 1995. One 1998 study argued that the Series had resulted in a lowest common denominator approach to voluntary environmental protection outside the regulatory framework: the Series was simply not strong enough to be used as an international tool for compliance; as it was too weak to be used for effective regulation, it should only be used only as a management tool.[38] Nevertheless there is some evidence that the adoption of an EMS can be linked to better site environmental management and regulatory performance, and can reduce air emissions and hazardous waste production.[39] More sceptical commentators argue that measuring environmental performance cannot be achieved using solely systems-based auditing, either ISO 14001, or EMAS. They provide evidence that an examination of 280 European companies at 430 locations provided no meaningful connection between certification and improved environmental performance.[40] ISO 14001 was a perfect tool for 'greenwash': even legal compliance could not be taken for granted.[41]

NGO and business informants looked at the Series' problem-solving capacity from different perspectives. There were a number of objections from NGOs regarding the value of putting an ISO EMS in place. The first was that the systems-based approach to environmental management was questionable conceptually. Because it had no performance requirements, it was untrue to claim that it was a system of environmental excellence and good environmental performance. The standards bodies and their dual imperatives of selling standards and appeasing business were the main reason for this weakness. Secondly, the preponderance of technical interests within the ISO system was seen as a barrier to solving specifically environmental problems. They also questioned the extent to which the ISO 14000 Series solved the targeted problems of pollution prevention and waste reduction, pointing to European studies, which were equivocal regarding the system's abilities in these areas. Business interviewees and informants from group three also identified a number of issues relating to the Series' actual contribution to environmental problem-solving. One obstacle resided with the standards bodies themselves. Because they were generally only responsible for setting standards they were less interested in how they were implemented. Secondly, the scheme only measured the extent to which a given management system complied with the standard; it did not demonstrate how the standard had improved the situation on the ground. The deliverables had been oversold, and had created problems with clients when they found out what they had actually bought. Because of its system-, not performance orientation it needed to be regarded as a management approach, not a marketing tool. One business interviewee commented that one of the Series' earliest and most influential industrial advocates had told them that the failure to include any performance requirements had been a big disappointment.

ISO 14001 has the prevention of pollution as a policy commitment. There is, however, a considerable difference between a policy commitment and an actual impact on environmental quality. ISO 14001's weakness with regard to the latter is attributable to the fact that the standard sets no thresholds, and contains no specific performance targets. These aspects of the standard have been visited several times in the course of this study, and they appear continually in the literature, as well as amongst some of the key informants interviewed. The ISO 14000 Series has not been highly successful in tackling pollution on a global level, but it has made a contribution. Had the interests behind the creation of the SAGE and TC 207 made a different set of decisions regarding the future ISO 14000 Series, the impact may have been greater. Given the equivocal nature of the literature and the absence of performance measures necessary to render the standard effective, the score is **low**.

Durability

In the past ISO's activities have been oriented around producing technical standards for specific products, but the more recent expansion in its scope is having wider impacts on social and environmental policy generally. These activities include environmental and water management, greenhouse gas emission accounting and verification, and the ethical behaviour of business, now the subject of an international standard for social responsibility.[42] TC 207 claims it has also adapted to the need for greater NGO inclusion, attributing this to the 'flexibility and latitude' that ISO's committees and sub-committees enjoy under ISO procedures.[43]

Interviewees from all three groups acknowledged ISO's expertise. One of the organisation's strengths was that it has been developing standards for more than 50 years, and as a consequence it has an excellent skills base. A weakness was its historical association with largely technical standards, the procedures of which were less adequate for its new move into social responsibility and environmental management. However, there were other institutions that were better adapted to dealing with environmental issues and ISO's procedures were not really suited to making what amounted to political decisions about what was and was not acceptable. NGOs interviewees were ambivalent about ISO's adaptability. One commented that there were now large questions hanging over TC 207 in the wake of the CAG–NGO task force collapse. At the same time, however, the innovations in procedure adopted by the Working Group on Social Responsibility were noted. These had created a precedent that national standards bodies, which were generally quite inflexible, would have to follow eventually. They predicted that at some future date standards bodies might be asked to change their procedures, and since they had changed them in the past there was no reason not to do so in the future. Another interviewee pointed to a lack of a culture of political debate as an impediment to change. They saw part of the problem as being linked to the fact that if ISO's standards setting procedures were ever questioned, this was interpreted as criticism. Interviewees from the third group identified the need for ISO to change, particularly with regard to how it managed interest representation. ISO was struggling to deal with the different needs of international interests and national member bodies. Change was essential or else the organisation would be left behind. But it was currently unclear how much its members were prepared to embrace new ways of doing things. Nevertheless, the institution was changing, albeit slowly.

ISO has demonstrated a degree of adaptability in moving first from the standardisation of products, to processes, and now systems. It is also increasingly active in more political arenas, including those related to environmental sustainability and social responsibility, and has demonstrated a degree of flexibility to accommodate these new interests in decision-making. The arrival of new stakeholders onto the scene as a result has been generally accepted by ISO itself, but there is also a marked degree of resistance from national standards bodies. Here, there is clearly an attempt to resist the 'creep' of NGO power into their traditional domain. This resistance mediates an otherwise very high rating to **high**.

Conclusions

ISO has reached a significant point in its history, but where it goes from here is uncertain. It has moved beyond the development of product standards relating to such relatively simple concepts as the orientation of the thread on a screw, the subject of its first standard. The change from product to process standards in the ISO 9000 Quality Management Series marked the transition. Since Rio, with the support of world governments and WTO, it has been elevated to a leadership position in such highly complex global issue areas as environmental management and social responsibility. Previously, ISO merely provided a decision-making space for deliberating technical solutions in response to the needs of a client base largely comprised of business, or manufacturing, interests. Now it is obliged to cater to the economic, environmental and social dynamics of sustainable development, which include balancing the demands of a range of multi-stakeholders operating at multiple levels.

Whether ISO can, or will, change at a more profound level remains to be seen. The problems associated with moving away from its existing structures and processes, not the least of which is the one-country-one-vote approach to democracy, are likely to make any change extremely slow. Some of the larger issues, such as the enfranchisement of civil society, or North/South equity, whilst acknowledged, continue to meet resistance from some national standards bodies. This resistance would appear to highlight the tensions between state and non-state interests in the ISO system, or perhaps better stated, between those interests currently benefiting from the status quo, and those seeking more influence.

TC 207's decision to pursue a systems- rather than performance-based approach to environmental management was to become a major sticking point for ISO's new environmental and social stakeholders. They objected both to the Series' failure to embrace any performance targets and the institution's exclusive participatory arrangements, in which environmental and social stakeholders were effectively second-class citizens. These two elements continue to form the basis of civil society criticisms of the Series. These perceptions were reinforced by the rejection of the reforming proposals of the NGO-CAG task force. The exact reason for the collapse of this initiative is difficult to explain. The crucial event appears to have been the NGO implication that if ISO was to move into more environmental social standards, it should follow those bodies with greater expertise, such as ISEAL. This seems to have offended some member bodies; they may have felt that 'standards are standards' (whether they are social, environmental or technical) and national procedures were already adequate. Here, there seems to be a basic clash of values. However, the fact that TC 207 was itself prepared to examine this issue, that the Working Group on Social Responsibility has partially redefined interest representation, and that the TMB is watching this process, demonstrates that the understanding of participation is in a state of flux across the ISO 'family'.

ISO 14001 is intended to encourage organisations to commit themselves to compliance with relevant domestic legislation and regulations, continually improving their overall environmental performance and preventing pollution. However, with respect to environmental performance it is the company that determines the level it wishes to reach and what objectives it intends to meet. The extent to which the ISO 14000 Series can be shown to have improved environmental performance and to have tackled the problem of industrial pollution is open to question.

5
Programme for the Endorsement of Forest Certification Schemes

Historical overview

Origins, institutional development and controversies

The creation of PEFC can be linked to fears amongst certain forestry interests of a monopoly of FSC certification in Europe. German publishers and printing houses had been under pressure from Greenpeace since 1993 to cease using paper sourced from old growth or poorly managed forests in Nordic countries. In 1995 the Association of German Paper Producers and the Association of German Magazine Publishers issued a statement committing themselves to products that could be sourced under a credible global programme. NGOs had not insisted on FSC, but publishers had the FSC programme in mind. In cooperation with various European forest owner groups, Finnish forestry interests collaborated in a campaign focused on Axel Springer Verlag as well as Otto Versand, two of the largest publishers and consumers of FSC paper in Germany. In December 1997 800 people from 11 European countries demonstrated outside the companies' offices in Hamburg, condemning the companies' support for FSC. This event has been identified as 'the psychological point of origin of the European forest certification system'.[1]

By the late 1990s European forest industries began to see the need to have 'credibility without tight engagement with the FSC' and to pursue their own certification initiative via an alternative market system.[2] European efforts were again led by Scandinavia, and between 1995 and 1996 forest industries and forest owners in Norway, Sweden and Finland attempted to develop a 'pan-Nordic' project. This foundered on account of NGO opposition to anything other than FSC certification. Forestry interests, on the other hand, preferred national certification initiatives, which they considered more cost-effective, and better suited to small-scale forest ownership structures, although it was understood that they were not sufficiently credible for non-domestic markets. The view emerged that a pan-European system, into which national schemes could fit, would provide the necessary international framework for forest certification. By this stage, the conclusion had been reached that the FSC was not meeting the needs of private forest landowners, and Scandinavian landowner associations in particular were already feeling alienated from FSC processes.

In 1998, the Finnish Forest Industries' Federation (FFIF), having failed to successfully engage with NGOs in the development of its own Finnish Forest Certification Scheme (FFCS), raised the idea of enlisting the support of governments to develop an EU-wide certification and labelling initiative. A meeting was held between the forestry ministers of Finland and Germany in March. German and Finnish forestry sectors, previously ambivalent, if not hostile to certification, agreed that a European certification system should be developed, mainly to develop 'a functioning European system before the FSC gained a foothold'.[3] A decision was made to keep government participation in the background, but the state authorities of Finland, Germany and France provided behind the scenes support. Attendees at foundational meetings at this time consisted largely of forest owner associations, industry interests and experts from Europe, although representatives from the American Tree Farm System, Australia and Malaysia also attended as observers. Between 1998 and early 1999 the idea that the scheme should not just be confined to Europe was given serious consideration as this might be perceived as constituting a barrier to trade; and a solution might be to initiate a process of mutual recognition for national certification programmes beyond Europe.

How to include dissenting interests therefore became a major issue for the emergent PEFC scheme, particularly if participating countries had already developed their standard. At an August 1998 meeting in Helsinki, it was agreed that environmental NGOs were not be invited to participate until forest interests had determined the framework of the scheme, and only then would they be consulted. This only seemed to confirm NGO suspicions about the system. The decision was to cause problems, as one of the guiding principles of the system was that it should be developed by means of open access and 'participation that seeks to involve all relevant parties'.[4] Since this had not occurred, it was necessary to clarify that the scheme would be a 'voluntary private sector initiative based on a broad view among relevant interested parties'.[5]

It was also agreed that the perspectives of other parties (presumably meaning NGOs) would be considered, but that they would be informed subsequently about the process. Only then would they be invited to participate in subsequent events.

Consequently, even before its launch PEFC was being criticised by NGOs as an attempt by European forest industry groups to create an international FSC-alternative. A major concern was that as a membership organisation, PEFC would only give voting rights to national governing bodies established by forest owners and industry associations, who were not obliged to include other interested parties. NGOs also pointed out, quoting from the scheme's own documentation, that the forest owners themselves were responsible for the development of forest management criteria. This made it clear, it was argued, that the PEFC was an organisation created by and for forest owners. It would also not enjoy much support from European NGOs unless it developed performance criteria, and although European NGOs acknowledged forest owners' problems with the FSC, they argued that their needs would be better served by engaging in dialogue with FSC.

In June 1999 the scheme was launched in Paris under the name Pan-European Forest Certification (also PEFC). 11 country representatives from Austria, Belgium, the Czech Republic, France, Finland, Ireland, Norway, Portugal, Spain, Sweden and Switzerland participated. The first PEFC schemes from Finland, Norway, Sweden, Germany and Austria were endorsed in 2000 and in May 2001 the Canadian Standards Association (CSA) joined the PEFC Council (PEFCC). Canada's certification scheme, known as the National Standard for SFM (Z809), had been in operation domestically since 1996. PEFC passed the FSC's area of certified forest for the first time in July 2001 (36.42 million hectares to 22.38 million hectares). The endorsement of the French, Latvian, Czech and Swiss schemes occurred in the same year, followed by Spain and Denmark in 2002, and in the same year the revision process began for some of PEFC's earlier standards, most notably in Finland and Norway.

Shortly after the launch of the scheme German environmental NGOs warned printers not use PEFC certified paper. They pointed in particular to the scheme's failure to ban clearfelling and the use of pesticides, the shortcomings in monitoring provisions, and the minimal ability for environmental and social groups to exercise any influence. By February 2001 relations between NGOs and PEFC had deteriorated to such an extent that they were being described as engaging in a 'tribal war'.[6] Most NGOs seem to have displayed their displeasure with the PEFC system by refusing to participate in standards-setting processes. NGO engagement with PEFC was consequently largely confined to the production of detailed critiques outlining the structural and procedural shortcomings across the system, and in comparative studies with other schemes. PEFC members and supporters roundly condemned the studies. When the WWF challenged the legality of the Austrian scheme, PEFC responded by accusing WWF of issuing 'inaccurate, sensationalist...and...discredited' information.[7] It asserted that: 'WWF only supports the FSC labelling system, which was originally designed for tropical rainforests'.[8]

However, environmental NGOs did not constitute an entirely united bloc of resistance to PEFC. At the organisation's General Assembly in July 2001 two individuals from Germany and France who declared an environmental interest were elected to the Board of Directors. Although both had extensive backgrounds in forestry and also served on the national Boards of PEFC, they nevertheless also represented national environmental organisations in their own countries. Other environmental groups preferred to view the participation of the French peak environmental organisation France Nature Environnement (FNE) as an exception to the general practice of NGO non-participation. In the case of Germany, NGO critics dismissed the groups as 'small environmental and social NGOs'.[9] PEFC Germany commented that it was correct that major NGOs were not participating in the scheme, but claimed that 'this is not the fault of the system but rather it is the intention of these groups to weaken the PEFC process by their absenteeism', suggesting that: 'ENGOs should carefully consider if it is worth to continue [sic] the struggle for power, or if it would be wiser to participate in a system that is already delivering on 40 per cent of the German forest area.'[10]

Despite these incremental gains in national NGO support, a somewhat frustrated note can nevertheless be detected in PEFC literature during 2001. PEFC stressed that NGOs could participate in national PEFC institutions and vote in them. WWF's accusations that the Austrian system had no minimum ecological standards and insubstantial controls and that that environmental organisations had no influence on PEFCC were dismissed as being 'substantially wrong' and constituted an 'attempt to shame the successful PEFC Initiative'.[11] Differences between PEFC and NGO supporters of the FSC were not always irreconcilable, however. Discussions in Sweden between PEFC and FSC interests resulted in an arrangement for the processing of timber by mills certified under either programme, the so-called 'Stock-Dove' process. Nevertheless, NGO objections also began to emerge elsewhere in the PEFC system, most notably Malaysia, Chile and Australia. These national schemes were accused by local NGOs of being dominated by industry, imbalanced in their arrangements for stakeholder participation, and lacking in adequate performance-based thresholds.

Towards the end of 2002 another four national schemes from Australia (Australian Forestry Standard Ltd – AFS), Brazil (Sistema Brazileiro de Certificação Florestal – CERFLOR), Chile (Certificación Forestal en Chile – CERTFOR) and Malaysia (MTCC) were elected at the General Assembly as PEFC members. The recruitment of such schemes into the PEFC camp has been linked to more concerted efforts to go global. As a result, the General Assembly became the main forum for discussions regarding the globalised format for the system during this period. These discussions were followed by an invitation from Gabon to the PEFC Chairperson to discuss the feasibility of developing a pan-African forest certification system. The possibility of expansion led the Board of Directors to establish a Globalisation Working Group. Determining to keep the same initials but recognising that it had broadened its geographical scope over the previous four years, the General Assembly agreed to change the scheme's name and orientation in October 2003.

PEFC was consequently renamed the Programme for the Endorsement of Forest Certification Schemes in the same year. Accepting the importance of global public perception, and recognising the need to gain support from NGOs, PEFC France and French NGO FNE organised a symposium in France in May 2004 for environmental NGOs interested in and supportive of the PEFC approach. The purpose was to bring about an exchange of views between participating environmental NGOs regarding the system, and to discuss creating an international NGO network to express their views internationally. Fifteen NGOs from 13 countries attended. In an attempt to consolidate national NGO support, agreement was reached to establish an informal global network built around national and sub-national NGO interests involved in national schemes. FNE was given a leading role in the new network and took on the secretariat of what was referred to as an international platform, 'open to all international, national and sub-national ENGOs wishing to take part in the ongoing development of PEFC national schemes'.[12]

The scheme's global expansion continued in 2004 with the launch of a promotional campaign in Asia, sponsored by Australia's National Association of Forest Industries, one of the original instigators of the Australian scheme. The Australian, Italian, Chilean and Portuguese national programmes were also endorsed as member schemes. In 2005, the certification schemes of Canada, Brazil, Luxembourg and the Slovak republic were also formally endorsed as well as the US-based SFI. With this growth in international membership there also came a shift in governmental attitudes towards PEFC, and from this point onwards, European markets – and government procurement policies – began a process of mutual recognition of both FSC and PEFC schemes.

The conflict between WWF and PEFC took a new turn in 2004. PEFCC sent an open letter to the World Bank/WWF Global Forest Alliance challenging the lack of multi-stakeholder participation in the development of a questionnaire assessing the comprehensiveness of PEFC and FSC forest certification schemes in Europe. The questionnaire was condemned as having a strong bias towards FSC, and PEFC called upon the Alliance to drop what is saw as flawed methodology. Instead it offered an alternative process for 'open and transparent discussion... that all stakeholders [could] have trust in'.[13] PEFC affiliates also came into conflict with the World Bank on a separate, but related certification matter, as the Bank's forest certification fund had refused to provide support for PEFC certification on account of its lack of inclusion of indigenous people, particularly the Sámi of northern Scandinavia.[14] These affiliates criticised this policy as being inappropriate to the evaluation of sustainability in Europe.[15]

World Bank problems notwithstanding, PEFC worked to ensure external international recognition throughout 2004. One important development in this direction was its acceptance as an association member by the International Accreditation Forum (IAF). This alliance was identified by PEFC as an opportunity 'to ensure that accredited forest certification certificates issued in one part of the world are recognised everywhere else'.[16] At the same time it also gained consultative status in the United Nation's Economic and Social Council.

In the latter half of 2006 PEFC received a further boost to its credibility when its social criteria were recognised, along with those of FSC, as 'good' in a test of ten eco-labels undertaken for German trade union IG BAU. PEFC was particularly recognised for its requirement that companies in the system offer training places. 2006 was also a significant year for the expansion of the PEFC brand. Asia and Africa experienced an expansion of the PEFC's CoC certificates. Other countries identified with PEFC in its global statistics for this year include Hungary, Japan, Morocco and the Netherlands along with Poland and Slovenia, whose schemes were under assessment. PEFC members lining up for, or pending, endorsement included the UK, Estonia, Malaysia, Russia, Gabon and Belarus.[17] By the end of the year, thanks to its endorsement of a number of previously discrete schemes such as the SFI, PEFC controlled 69 per cent of the world's certified timber, in comparison to FSC's 24 per cent. As of 2007 PEFC had a membership of 33 NFPs, 22 of which had already undergone assessment, with the remaining country schemes being at various stages of development and recognition under PEFC procedures. PEFC's area of certified forests rose from 32.37 million hectares in 2000 to 196.325 million hectares by April 2007. CoC certificate holders increased from approximately 100 in 2001 to 3127 by April 2007.

PEFC's efforts to enhance its public image continued in 2007. In an effort to provide the facts 'behind a number of incorrect assertions' concerning indigenous peoples' participation, environmental NGO support for PEFC and its social dialogue, PEFC issued a brochure claiming it was supported by environmental NGOs in April of the same year.[18] Both French-based FNE, and US-based Conservation International (CI) were also listed as supporters. The participation of the Austrian organisation Umweltdachverband, representing 34 national ENGOs, was cited as evidence of support of PEFC by environmental NGOs. Towards the end of year it was also granted observer status in the MCPFE, the European intergovernmental forest process.

In October PEFC resolved to undertake a comprehensive review of its governance.[19] A panel of forest experts was convened, and included three members of the Board of Directors and four independent individuals (two of whom were from international environmental and animal welfare organisations). The mandate of the panel was to review PEFC's effectiveness with regard to those aspects of its governance concerning rules, standards, implementation and monitoring, the extent of support from existing members, and to engage in dialogue with environmental NGOs.[20] The review followed the general analytical methodology of the international institution One World Trust, which evaluates quality of governance based on three governance attributes: participation, transparency and responsiveness.[21] In terms of participation, the review recommended that PEFC should develop a stakeholder engagement policy ensuring meaningful participation. Specific elements of recommended action included the convening of a PEFC-sponsored conference to initiate dialogue with environmental NGOs, identifying and recruiting local stakeholders to participate in national governing bodies and providing means for them to participate substantively in PEFC at the international level. To this end the review proposed the expansion of the existing extraordinary members into a broader, forum-based range of stakeholders, to be given rights and responsibilities in the organisation's governance. The resulting Stakeholder Forum would be permitted to nominate two positions of the Board of Directors, and to the PEFC's Panel of Experts. The Forum would be enfranchised with a one-member-one-vote method of reaching agreement, and be given powers constituting a maximum of 50 per cent of the votes allocated to PEFCC itself. Measures for improving the transparency of the institution were also identified as a priority for the institution. The review concluded that PEFC's existing requirements for dispute settlement were adequate, but this was somewhat mitigated by the recognition that PEFC had no control over the existing complaints and appeals procedures of its associated organisations. It could also help direct complaints to the appropriate point in the system to make PEFC's existing decentralised system more user-friendly.

Other elements of note within the review included a recognition that the PEFC needed to respond to criticisms more constructively, under the maxim 'the more strident or vitriolic the attack, the more gracious and engaging the response'.[22] It was also recommended that an initiative aimed at creating new alliances and advancing PEFC's standards for SFM in the tropics should also be developed. A technical unit should also be established to provide capacity-building assistance to members and prospective members. Substantial changes were also recommended relating to the institution's organisational structure at the international level, as well as its assessment, endorsement and mutual recognition processes.

PEFC invited WWF join the proposed Stakeholder Forum. WWF responded in February 2009 by noting that the review had put some important steps in place to improve the governance of the system. But it wanted to see balanced representation throughout all levels of the system, definitions of 'controversial' timber sources closer to its own understanding, greater consistency between national schemes, and public summaries of certification reports (all of which it claimed FSC delivered and PEFC did not). It declined to participate in the Forum, arguing that the 30 per cent power the Forum would hold in the General Assembly still did not prevent PEFC from being 'dominated by a single interest group'.[23] PEFC responded in March by calling on WWF to reconsider its decision, but concluded that it seemed unlikely either party's views would ever be fully aligned. It questioned WWF's decision and rejected its assessment of the merits of FSC, based as they were on the forest certification assessment criteria developed as part of the

WWF/World Bank Global Forest Alliance. WWF's support for FSC, and the decision not to participate in the Stakeholder Forum, it argued, were 'based on a report of questionable rigour and objectivity'.[24]

Governance within PEFC

System participants

Participation within the PEFC system occurs both internally and externally on several different levels (internationally, nationally and/or regionally), and within a range of processes associated with standards-setting, endorsement and certification. Membership consists of national governing bodies, whose objective is to instigate the creation of a PEFC scheme within their own country, and who may apply to join the international body, referred to as the PEFCC. Following ISO 14004 the Council identifies a stakeholder or interested party as 'an individual or group of individuals with a common interest, concerned with or affected by the operation of an organisation'.[25] Other associations, such as international organisations, may become members, but they are classified as extraordinary members with no voting rights. By the end of 2006 there were 11 extraordinary members representing forest owners, managers, processors and trading interests. PEFC's Statutes are not explicit regarding the membership composition of the national governing bodies, which have sole responsibility for determining who should be invited to become members. These provisions have resulted in a degree of variation in stakeholder composition in national governing bodies.

Institutional arrangements

International level

PEFC is based in Luxembourg, and describes itself as 'a global umbrella organisation for the assessment and mutual recognition of national forest certification schemes developed in a multi-stakeholder process'.[26] See Figure 5.1 below.

The name PEFCC is used to describe the associative relationship between members of the institution. The PEFCC is the governing body of the scheme, and the formal representative of the system with a minimum size of six members. The General Assembly is the highest body of the PEFCC. It meets annually and comprises one official delegate per national programme, who may be accompanied by up to two observers from its governing body. The assembly has

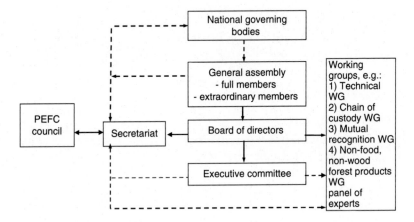

Figure 5.1   Structure of PEFC.
Source: PEFC Technical Document, p. 5 (adapted); solid black lines on arrows indicate entities associated with PEFC Council.

a number of tasks. It elects and dismisses the Board of Directors and members, revises the Statutes, documents and procedures of the scheme, adopts the budget and accounts and selects the auditors, chooses the institution's location and is responsible for dissolving the PEFCC. It also elects the organisation's Chairperson, and First and Second Vice-Chairperson.

The Board of Directors consists of the Chairperson of the PEFCC, two Vice-Chairpersons and two to ten members. The Directors of the Board are the administrators and managers of the PEFCC. The Board's more substantive activities include promoting the scheme and undertaking public relations initiatives, as well as determining the conformity of applicant schemes to PEFC requirements, and considering the suitability of non-PEFC schemes for mutual recognition. The Secretary General is answerable to the Board, and exists to ensure communication between members, support the work of the Board, attend meetings, take minutes and manage the Secretariat, which can be based in any of its member bodies. An Executive Committee, consisting of the Chairperson and the Vice-Chairpersons, and permitted to co-opt other Board members as required, exists to undertake tasks assigned it by the Board.

National level

The normative operational bases for PEFC-member schemes are: the PEOLG; the Principles, C&I for the SFM of African Natural Tropics

(ATTO/ITTO PC&I); C&I for the Conservation and Sustainable Management of Temperate and Boreal Forests (Montréal Process C&I); and a range of other regional processes. These existed prior to the establishment of PEFC. National schemes are instigated by 'national forest owners' organisations or national forestry sector organisations having the support of the major forest owners' associations'.[27] Figure 5.2 below outlines the institutional relationship between national governing bodies, the PEFCC and other related entities.

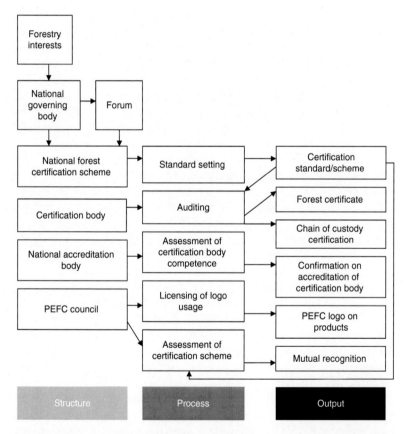

Figure 5.2 Structural and procedural arrangements for standards-setting and related activities within the PEFC system.

Sources: http://www.pefc.org/internet/html/acitivities.htm (adapted); Rules for Standard Setting, p. 3, PEFC Technical Document, p. 6.

At the behest of this national governing body of forestry organisations, interested parties are subsequently invited to constitute a forum
for the development of national standards. Such parties 'should' represent the 'different aspects of sustainable forest management' and
include forest owners, forest industry, environmental and social NGOs,
trade unions, retailers and other relevant organisations at a national or
sub-national level.[28]

Standards development, accreditation, chain of custody and certification

PEFC's requirement for the setting of standards follow ISO Guide 59 and
specifies that 'all relevant interested parties shall be invited to participate
in the standards-setting and the PEFC principles on transparency and
consultations shall be respected'.[29] National forums are responsible for
standards-setting and constitute committees, councils, working groups
or similar. Formal consultation regarding the drafting of standards consists of the views of interested parties arising from consultation being
'discussed', and the national forums must give 'general information'
on the changes made in the light of the consultation process.[30] Once
national standards and national schemes have been developed, there
is a process for their endorsement, mutual recognition and revision by
PEFC. The process of assessment is 'transparent and consultative', and
occurs on international and national levels.[31] Procedures also exist for
the development of CoC standards, although it is not clear if there is
any public participation.

Certification enterprises must also be accredited by a national
accreditation body, which in turn must be a member of either the IAF
or the European Cooperation for Accreditation (EA), follow the requirements of ISO Guide 61 and be familiar with ISO quality and EMS (ISO
9000:2000 or ISO 14001) or the EMAS. As of 2005, certification bodies
that had accredited forest management or chain of custody standards
without these necessary provisions were given four years – from the
issuance of the first national scheme certificate – to catch up with
accreditation requirements.[32]

Certification under PEFC covers forest management on a regional,
group and individual level and CoC. Forest management certification under the PEFC framework is based on existing international
management systems, product certification and the EMAS verification procedures. Management systems certification can be either quality or environmental, and all procedures are based on either the
relevant ISO Guide (62, 66, 65) or EC Regulation 761/2001, as

appropriate. Additional procedures covering forest management and CoC certification may be developed by the given scheme, but this is optional. In 2002, the General Assembly also agreed to the development of an international CoC programme. Building on the work of the Confederation of European Paper Industries (CEPI) and the European Confederation of Woodworking Industries (CEI Bois), the standard was approved in 2004.

Institutional typology

PEFC relies heavily on European and other regional intergovernmental forest policy frameworks and comprises national certification programmes generally recognised by national governments. National governments have also played – and continue to play – an important role in its development. For example, governmental forest agency representatives have sat on the Board of Directors. Clearly, non-state interests constitute the most significant players in the system, but as these largely consist of economic interests, it is an exaggerated claim to assert that PEFC represents a 'pure' non-state system comprised of a wide range of non-state actors. It belongs in the state-centric, rather than the non-state end of the continuum, but given its relatively autonomous nature, with a rating somewhere between **medium** and **low**.

Following normative international trends, PEFC lays claim to being consensus-based in standards-setting. However, once they are submitted for endorsement internationally, standards require a qualified majority vote of two thirds of the General Assembly. Furthermore, in both the General Assembly and the Board of Directors, the PEFCC functions democratically on the basis of voting by simple majority. These factors place the institution towards the aggregative rather than the deliberative end of the democratic continuum, but with a rating of somewhere between **medium** and **low**.

As a network of largely autonomous, self-governing national schemes embedded within national regulatory frameworks, PEFC is relatively 'new' in its governance style. The institution can also be interpreted as constituting a series of public-private partnerships between governments and forest interests. These considerations also make it relatively innovative. However, the close linkages from sub-national/national programmes, upwards to intergovernmental forest policy frameworks tie it to more classical, hierarchical notions of governance. This places it on the 'new' end of the innovation continuum, but with a rating of **medium**. See Figure 5.3 below.

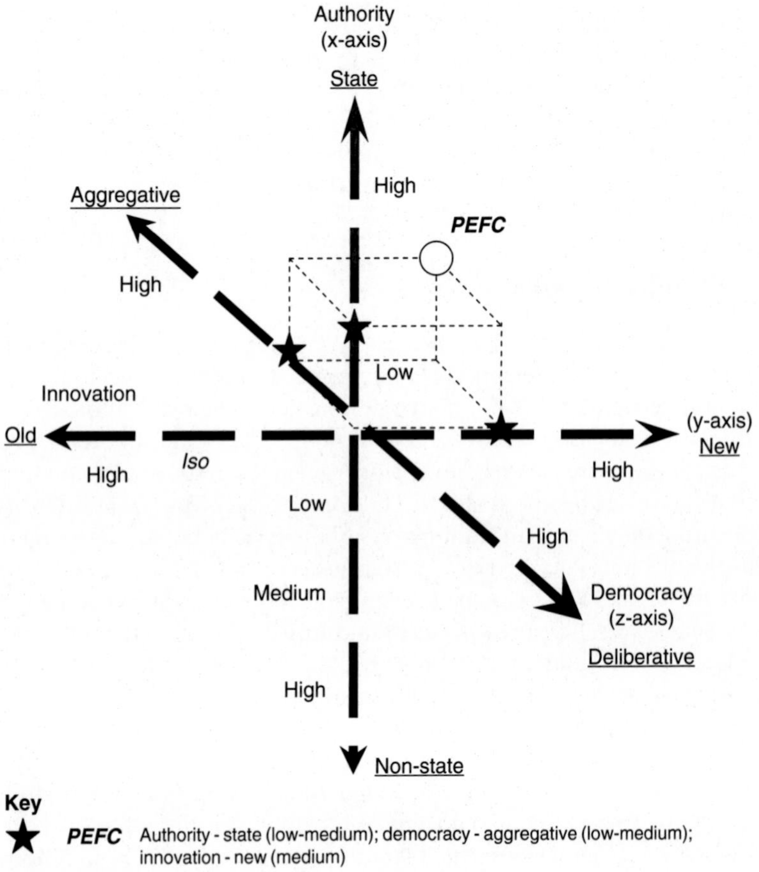

Figure 5.3 Institutional classification of PEFC.

Governance quality of PEFC

Commentary

PEFC received 26 points out of a maximum total of 55. Seven indicators achieved low ratings (inclusiveness, equality, resources, accountability, transparency, democracy and problem-solving), and four medium ratings (agreement, behaviour change, dispute settlement and durability). There were no very low scores. The conventional pass/fail target value of 50 per cent was met by two criteria (decision-making and implementation). See Table 5.1 below. The scores achieved were 53 per cent (implementation and decision-making), and 40 per cent (interest

Table 5.1 Evaluative matrix of PEFC governance quality

Principle: 1. Meaningful Participation

Criterion	1. Interest representation Highest possible score: 15 Lowest possible score: 3 Actual score: 6			2. Organisational responsibility Highest possible score: 10 Lowest possible score: 2 Actual score: 4		Sub-total (out of 25): 10
Indicator	Inclusiveness	Equality	Resources	Accountability	Transparency	
Very high						
High						
Medium						
Low	2	2	2	2	2	
Very low						

Principle: 2. Productive deliberation

Criterion	3. Decision-making Highest possible score: 15 Lowest possible score: 3 Actual score: 8			4. Implementation Highest possible score: 15 Lowest possible score: 3 Actual score: 8		Sub-total (out of 30): 16
Indicator	Democracy	Agreement	Dispute settlement	Behaviour change	Problem-solving	Durability
Very high						
High						
Medium		3	3	3		3
Low	2				2	
Very low						
Total (Out of 55)						26

representation and organisational responsibility). At the principle level, the aggregate result was 40 per cent for meaningful participation, constituting a failure to reach the target value of 50 per cent. The aggregate result for productive deliberation was 53 per cent, thus reaching the target value.

Evaluation

Interest representation

Inclusiveness

Unhindered participation is seen as a fundamental concept of sustainable governance.[33] A general trend has also been identified in forest certification towards greater participation, which poses a dilemma for the forest industry when it comes to improving, or maintaining, market access through initiatives such as certification. It needs to ensure inclusiveness, which delivers credibility, but also needs to respect forest sector sensitivities, thus ensuring producer participation. The approach adopted by schemes dominated by forestry interests such as PEFC may in fact reflect a contending conception of participation, but one that is nevertheless still subjected to an 'upwards pull' towards greater participation.[34] The voluntary nature of inclusiveness arrangements allows PEFC considerable latitude regarding interest representation and forest owners and industrial interests predominate. The institution itself states that 'the participatory elements in the PEFC framework are applied predominantly at the national and sub-national level'.[35] Here, it is only a recommendation, and not mandatory, that all stakeholders should have say in the decisions made by the national governing body.[36]

Forestry interests are responsible for inviting 'national organisations representing all relevant interested parties' to form the basis for a PEFC national governing body.[37] Country schemes have their own preferences regarding which stakeholders should participate in standards-setting. This can include government. Claims that national standards have been approved with the participation of major NGOs have been challenged by NGO critics as a 'misrepresentation of reality', and on the grounds that 'there is very little proven support for the PEFC in any of the countries which either have, or are developing standards and/or schemes'.[38] Participation in certification assessments is confined largely to the auditing phase of the certification process. At this point the certifier is obliged to 'include relevant information from external parties (e.g. government agencies, community groups, conservations organisations [sic], etc.), as appropriate'.[39]

NGO interviewees characterised their involvement in PEFC as 'non-existent'. At an international level there was no room for NGOs to become part of the PEFCC other than as observers. The Board was only open to members of national governing bodies and even international organisations such as the ILO only enjoyed observer status in the Council. The structures of national certification schemes militated against any ability to influence their outcomes and it was as hard to have an influence now as in pre-PEFC days. It was wrong to describe PEFC as inclusive of environmental interests simply because it had some national environmental NGOs participating. A significant weakness of the system on the national level was the lack of stakeholder participation in both standards-setting and the actual certification process. This was reflected in the treatment of indigenous interests. Most business interests argued that NGOs were adequately included in the system, so long as it was understood that those involved were 'not as radical as the international ones', such WWF and Greenpeace. They accepted there was a preponderance of forest owner, forest industry and forest-dependent trade representatives within the system. But PEFC was essentially an initiative by the forest sector and forest owners. One interviewee identified private forest owners as being at the heart of PEFC and said they could be described as its 'political arm'. Another interviewee, from the third group of informants, was highly critical of the exclusive nature of PEFC in their local region:

> Once the scheme was more or less designed PEFC asked the regional government if it wanted to participate. I asked them who was part of their selective club, and whether environmental NGOs were there. They said yes, but when I asked which groups, it was clear that their idea of an environmental NGO were hunters, or people who wanted to go freshwater fishing. These kinds of organisation were already very close to private forest owners ideologically. I replied by saying that the way PEFC had organised itself was not participatory enough. The government wanted the discussion about SFM to be one, which was representative of all the different views on forests. You cannot just form a selective club with your own friends, and then leave others behind. It is not acceptable. So the government stayed out, and did not get involved. But PEFC went ahead and developed its own system anyway.

On an international level, the Board of Directors can claim to include social and environmental sectors, but they are outnumbered. The General Assembly contains an even weaker spread of interests, mostly attributable to its membership structure, which permits only national governing bodies to actively participate, but on behalf of the country schemes, not stakeholder sectors. On a national level, the PEFCC Statutes prevent any interest other than forest owners or forest industry initiating the development of the country scheme. The national governing bodies themselves, once formed, should represent diverse interests, but this is not mandatory. It is only at the level of standards-setting that the PEFCC Statutes oblige the national governing bodies to invite all relevant interested parties to participate. But even here it is unclear in this wording, if it is the act of inviting, rather than the participation itself that is mandatory. In terms of the certification processes themselves, interests are only represented in so far as they can provide information relevant to the assessment of conformity of forest management activities to the standard. PEFC's inclusiveness is **low**.

Equality

PEFC's imbalance in stakeholder representation, it is argued, provides limited opportunities for minority interests to influence rule-making within the system, particularly in terms of compliance and standard implementation.[40] As early as 1998 European NGOs indicated they could not support the emerging PEFC system if it did not make 'structural changes to ensure the equal input of other sectors'.[41] Not surprisingly, this has led environmental NGOs in particular to make a number of broader claims concerning the system.[42] It has been argued that, by ignoring environmental NGOs and other stakeholders, the scheme may have placed its reputation at risk.[43] Another major source of contention is the status accorded to indigenous interests within the PEFC system. After the withdrawal of environmental NGOs from the Finnish certification scheme, for example, the Sámi and state forest manager Metsähallitus came into conflict after the former claimed they had not been sufficently listened to in the process of defining the standard.[44] In this matter, the PEFC system as a whole fell foul of the World Bank, which would not permit PEFC to receive grants to expand its certification activities on account of its failure to acknowledge indigenous peoples' rights. PEFCC published a position paper regarding indigenous people in December 2005. The policy indicated that PEFC's recognition of intergovernmental arrangements, in combination with the member schemes' own national policy framework, provided for sufficient recognition and no further action was required.

NGOs universally condemned the lack of equality within the PEFC system. Unless the weighting of different interests was resolved, there was no point in participation, and their involvement would be misused, as had been borne out in some PEFC initiatives. They had objected to the disregard for indigenous peoples' rights when the scheme had started, and problems had subsequently escalated in Sweden and in Finland, where the differences between PEFC and FSC on this matter were substantial. As PEFC expanded into Southern countries, this weakness would become more apparent. Business interviewees saw PEFC as providing for greater opportunities to protect private property rights than FSC. They pointed to both the SFI and the AFS as being motivated by these considerations. For both large and small owners the decision to be involved in a given system all came down to the politics of the scheme. FSC did not deliver an equal say for economic interests. One interviewee from the third group of informants provided an alternative perspective. They felt that PEFC looked at things in an overly economic way. The system allowed for the maximisation of economic returns, but minimised the environmental effort required to gain such returns.

It is undeniable that forest owners and the forest industry played the most significant role in the establishment of PEFC. Many of the scholarly sources, as well as most of those stakeholders interviewed in this study, acknowledge the dominant role played by these interests within the system. Where other stakeholders do play a role within the system's participatory structures, it is limited. Forest owner- and forest industry interests initiate the development of national schemes, and even if other stakeholders are invited to participate in national governing bodies and standards-setting, it is on an unequal footing. This is particularly the case with indigenous peoples in the Nordic countries. Equality in the system is therefore **low**.

Resources

Information regarding PEFC's financial activities is not publicly available on its website, and no response was made to a request for financial information during the course of this study (see transparency below). According to one interviewee, the global budget in 2006 was roughly 500,000 Euros, which was expected to run the whole entity. Forest owner associations from supporting countries were identified as being one of PEFC's main funding sources. Another informant provided some detail regarding the financial arrangements associated with their national scheme. Funding had remained at a relatively constant level over the life of the scheme, and was derived from a forest industry fund, the purpose of which was to promote the use of wood. The body raised voluntary contributions from forest owners and wood buyers and had provided funds to the scheme annually. A very small amount of income, about 1 per cent, was generated from charging for the materials associated with the issue of PEFC logo licenses; this was not a user's fee, but covered the costs of producing CDs, documents, and so forth. In addition there were member fees levied on membership organisations. In terms of participation, the general rule was that stakeholders funded their own attendance, although there were some exceptions; some travel costs were provided to members far from the capital.

No participants interviewed received funding to take part in any of the system's standards-setting or governance processes. NGOs and business interests all confirmed they funded their own participation. A number of NGOs stressed the cost in terms of time and money of participating in PEFC either directly or indirectly. The expenditure of resources on PEFC had been at a level that was concerned mainly with analysing the PEFC's governance structures and providing external commentary. With limited other resources available major NGOs had determined that the time and effort required to engage with and influence PEFC was not worth the investment, particularly when efforts did not deliver anything. Business informants confirmed that there had been support from industry to underwrite PEFC. One interviewee commented that industry groups had provided technical support in the development of the PEFC's CoC standard, because it needed industry expertise. This was because this was industry's normal way of working; as a supporter of certification it could not keep out of the forest debate and just passively use the system. In addition to this direct support, industry had also had its own expenses in terms of time and money, such as attending PEFC General Assemblies, and hiring consultants. Another indicated their organisation had invested money and time in the PEFC process.

The evaluation of PEFC's provision of resources does not refer to its sources of general income, but rather the amount of funding, or general capacity-building it undertakes to encourage various interests to participate in the PEFC. As a rule, PEFC does not provide resources for the different sectors to participate internationally and there does not appear to be any mechanism to fund participation in PEFCC. Anecdotally, some key members are funded to participate in meetings at the national level, but more marginalised interests, such as indigenous people, must rely on support from external sources, rather than the institution itself, to have their interests represented. The provision of resources to participate in PEFC is **low**.

Organisational responsibility

Accountability

Although the General Assembly is identified as the highest body of the PEFCC, decisions that are 'binding to the PEFCC financially or politically' are the responsibility of the Board.[45] Accreditation procedures within the PEFC system occur on the national level. A problem identified with this approach is that national bodies accredit both systems and performance-based standards.[46] This can undermine the value of schemes within the system that use performance-based standards since a scheme, which may use the ISO 14000 Series, systems-based approach, such as Norway, can certify all national forests in one stroke, yet enjoy the same status within the PEFC as performance-based schemes. Furthermore the assessment of such region-wide certificates can be based on random sampling rather than comprehensive audits. There are also historical problems regarding the lack of public consultation and availability of public documentation associated with schemes that were accredited before 2005.[47] The decision to co-opt intergovernmental C&I as the underlying framework for PEFC also occurred before most interested parties had the opportunity to consented to their use. Environmental and social interests in particular had already expressed reservations regarding the content of the PEOLG even before they were adopted and adapted by PEFC.[48]

Several NGO interviewees expressed concerns that PEFC had used the involvement of their organisation as a means of endorsement, without actually taking their views into account. This had made them all extremely cautious about having anything to do with the system in the future, and in their dealings with those holding official positions. Business interests generally considered PEFC accountable. Individual interviewees from all three sectors expressed misgivings about regional or country level certification. Because PEFC certified by area and not one forest exclusively, the situation could arise where forest owners could end up being certified without knowing it.

PEFC seems to be more accountable 'vertically' to directly interested constituents internally than 'horisontally' to the general public at large, or to more 'peripheral' interests, such as indigenous people. This is problematic given the devolved methods of oversight whereby most responsibility for standard development and implementation is at the national level. Accountability within the PEFC system is considered to be **low.**

Transparency

On a national level, the written procedures of standards development forums must be available to interested parties if requested. Transparency within the context of standards-setting refers to communicating the commencement of the standards-setting process, and by making information on the development process available to interested parties through such media as the Internet. Standards must subsequently be reviewed in a 'fair and transparent' manner.[49] Some of PEFC's normative requirements, while emphasising transparency, are either vague regarding implementation, or restrictive in application. In the case of the former, whilst the views of all those who are invited to participate in the creation of national schemes and standards are expected to be noted, it is not specified if they must be addressed. Financial information is not publicly available.

Both national and international NGOs expressed concerns that the PEFC system was neither open nor transparent. The lack of public availability of documentation, the opacity of governance structures, and the unclear decision-making processes were the main areas of complaint. PEFC had made commitments to improving its consultation and participation processes, and its CoC arrangements, but the materials were not easily accessible or publicly available. This made it difficult to assess where PEFC stood on anything, or whether past failings had really been addressed. Business informants had various views regarding the transparency of the system. One, commenting on the transparency of the national standards-setting procedures noted that the standards were open for consultation, but another noted that the extent to which PEFC publicised its activities could be seen as a weakness of the system. Two informants from the third group did not believe PEFC needed to go to the levels of transparency required by FSC. The democratic mechanisms of society in general provided sufficient transparency in decision-making. However, another interviewee admitted that the lack of clarity regarding who was ultimately responsible for the approval and justification of decisions was a disadvantage of the system.

It could not be said that the initial development of PEFC occurred in a particularly open or transparent manner. This was recognised as a problem by some of PEFC's original members. Some information is not publicly available at all (such as financial matters), and much of what goes on inside the system's internal participatory structures, such as meetings of the Board of Directors and General Assemblies, continues to be opaque. This opacity may be partly attributable to the fact that transparency within the PEFC system has a particular definitional context, and refers more to the process of issuing public information than the openness of the system's structures and processes to public scrutiny. However, even in this more limited interpretation, PEFC has been criticised by supporters as well as opponents for failing to adequately publicise its activities. Information on some of the institution's activities is available through its newsletters and on its website; however, no response was made to a request for financial information during the course of this study.[50] Overall, transparency within PEFC is determined to be **low.**

Decision-making

Democracy

The observation has been made that 'PEFC relies on the democratic procedures where the society at large defines the rights and duties of different forest users and implements them through normative regulations'.[51] The notion that PEFC considers itself democratically legitimated by such external arrangements deserves some consideration in the light of how democracy is exercised internally. Within the General Assembly enfranchised members have between one and three votes commensurate with the annual cutting rates of less than 10 million m^3, between 10 and 30 million m^3, and more than 30 million m^3. This approach to decision-making has been criticised as rendering the system undemocratic, as it disproportionately favours large producers. This is compounded by the fact that forest interests also predominate at a Board level where decisions are reached on the basis of simple majority. Non-forestry interests are also in the minority in national decision-making bodies, and their involvement occurs at a relatively late stage of the process. The procedures adopted within PEFC have been defended by PEFC supporters, and contrasted with those used within the FSC, which has been accused of creating an 'artificial democracy' amongst participating interests.[52]

NGO informants considered it a waste of time to sit in so-called consensus processes, because the system was so weighted in favour of forest owners and the forest industry. In one national scheme, for example, although decisions were usually taken by consensus, this had proved unworkable for NGOs, because they were expected to support the national certification programme and its conception of forest management in order to participate. Business informants accepted that NGOs would be more likely to participate in PEFC if they were given greater voting rights, and that final decisions were made in 'rather small committees'. One informant from the third group of subjects questioned the fundamental validity of increasing NGO democratic rights. Environmental and social groups had that right in the FSC system, but this was not the case in the real world, because it was ultimately the owners who were responsible for undertaking forestry activities. The democratic processes within the nation-state inherently addressed the divergent views in society, and nothing more was required.

PEFC displays a clear institutional preference for leaving democracy within the confines of the nation-state. A common view amongst interests supportive of PEFC was that there were already sufficient safeguards in the governmental regulatory environment for stakeholders to influence decision-making regarding forest management. Internationally, when non-forestry interests are accorded the right to vote, the combined social and environmental forces are unable to outvote those representing economic interests. International voting procedures also mean that large forest producers outweigh smaller interests. The democratic legitimacy of PEFC is therefore open to question. The problem is not one of aggregative versus deliberative models, but rather that PEFC's democratic processes are not seen as neutral, and participants are treated differently, resulting in divergent levels of trust in decision-making authority. The opportunity for democracy to flourish in such an environment is **low.**

Agreement

Decisions in the General Assembly and the Board of Directors are generally made by a simple majority of votes. The Executive Committee, which sits under the Board of Directors, does not have any formal mechanisms for reaching agreement. Once they have been developed and assessed, national schemes are endorsed internationally by a vote requiring two-thirds majority firstly by the Board and subsequently by the General Assembly. It is interesting to note, in contrast to the majority voting associated with the PEFCC that the process of standards-setting at the national level is on the basis of consensus. National forums are obliged to develop their own written procedures for the formative stage of national standards development. Approval of the standards is also based on consensus. The definition of consensus within the PEFC system follows that of ISO. According to documentation, once the national body has developed the final draft of national standards, they are distributed nationally as part of a formal process of consultation.

The applicant national body requests comments from interested parties, which are passed on to the consultant. The PEFCC encourages everyone with an interest to provide comments to the consultant.[53]

Comments on how agreements were reached in the PEFC system concerned both voting and consensus, but had little to do with procedural mechanics, and more to do with politics. Several national-level NGOs commented that there was no consensus, and the process was always majority rule, although a further interviewee pointed to the Stock-Dove process in Sweden as a 'different dialogue' between PEFC and FSC interests, which represented something that the broader PEFC community could learn from. Several business interviewees commented on the nature of decision-making within PEFC and FSC. One argued that the structures of both PEFC and FSC were open to influence, and it was merely a question of the orientation and intention of the schemes. Decisions were made differently in each system, but in both cases the ground rules affected the nature of the decisions made. One interviewee argued PEFC's rules were a direct response to the shared decision-making processes within the FSC system. Another interviewee from the third group commented that the differences between private forest owners and NGOs as to how to reach agreement were both philosophical and political, and almost irreconcilable.

PEFC contains a range of procedures for reaching agreement. Generally speaking, PEFC functions on a majority voting system internationally. On a national level, there appears to be some variation in how national governing bodies function in terms of reaching agreement. There is also a lack of clarity on the exact definition and function of consensus within the different standards-setting processes of member schemes, and even if consensus is formally defined at the international level. An otherwise high degree of clarity at the international level regarding methods of agreement is offset by inconsistencies at the national level, resulting in a **medium** score.

Dispute settlement

According to PEFC, 'The right for appeals and appropriate grievance procedures related to the implementation of the certification schemes, ensures fair and impartial scheme implementation and certification.'[54] National stakeholder forums associated with standards development are expected to have written procedures outlining an appeal mechanism for impartially dealing with any procedural or substantive complaints that arise. The same procedures apply to developing CoC standards. Both the national governing entities and certification enterprises

are also required to develop dispute settlement bodies, or have procedures for developing such bodies on an ad hoc basis, to address
certification-related problems. Procedural inconsistencies make implementation difficult; in Finland, for example, the dissenting interests are
encouraged to comment, but there are no procedures for formally sub-
mitting complaints.[55] As part of the Governance review process PEFCC
adopted a procedure for investigating and resolving appeals at the international and national governing body levels in 2007. It does not cover
complaints regarding the activities of certification bodies, which are
to be dealt with under their own provisions, or via the IAF. Appeals
only cover decisions made by the Council's Board and Secretary General. Complaints and appeals are investigated via an ad hoc task force
group (TFG). The TFG submits a report to the Board, which either
approves or rejects the conclusions. Complainants are then informed
of the determination.[56]

NGOs interviewees were disappointed by the reality of dispute settlement versus written procedures. Concerns were also expressed regarding
the extent to which both the certification scheme and the certifiers
implemented the requirements for dispute settlement arising from
stakeholder consultation. One NGO reported that there had been so
many complaints raised against large forestry companies in their country by local communities and indigenous stakeholders that people now
considered that the companies were basically self-certifying. Such processes existed solely for PEFC to say it had a process. In reality forestry
interests did not want to know and simply disregarded these views.
According to the one business interviewee who did comment on dispute settlement, the view was expressed that conflicts between interests
were best resolved in the market. One interviewee from the third group
of subjects provided an interesting explanation as to why there were no
disputes within their country scheme. Conflicts were addressed through
the one-person-one-vote procedures of the Board and the National
Certification Council. Since NGOs had chosen not participate in the
development of the national system the problem of economic and environmental groups needing to resolve discussions by such means had
never arisen.

The settlement of disputes in a formal sense occurs only at the
national level within the PEFC system, although it is to be noted that the
recent governance review has recommended that this change. National
member schemes are obliged to create either dispute settlement bodies,
or have other procedures, but it is unclear how effective these mechanisms really are, or in some cases if they have been used at all. A further

problem may lie in the fact that complaints mechanisms are confined to matters regarding standards-setting, certification or implementation; broader conflicts have no venue. It seems more important in the PEFC system to be able to demonstrate the existence of dispute settlement mechanisms per se rather than have disputes settled. However, it should be noted that there is, as of 2007, a dispute settlement process under the direction of the Board of Directors. The ability of PEFC to settle disputes previously was low. It is unclear at this stage how effective these provisions are, but a provisional rating of **medium** is given.

Implementation

Behaviour change

Forest governance scholars have described PEFC as a 'strategic move to regain control over an issue area predominated by environmental interests and co-opt the discourse on forest certification'.[57] PEFC's framers were initially reluctant to adopt any C&I. However, as internal deliberations over the requirements for a certification system progressed, it was acknowledged that some guidelines were required as framework around which to develop national standards. However, the very value of using C&I processes as a basis for certified forest management activities is problematic. As early as 1997 in the IPF process it was recognised that 'C&I are not performance standards for certifying management'.[58]

NGOs also generally interpreted PEFC as a reaction, or as counterinitiative, to FSC. PEFC had muddied the waters and people were unable to tell the difference between the two schemes. This interviewee thought this strategy had been successful. The national systems that had joined PEFC, had arisen with the intention of avoiding the stricter demands of the FSC. NGO interviewees were prepared to concede that at least in terms of forest owners, attitudes had improved considerably from the early 1990s, and that attitudes regarding the merits of certification had changed. Improvements in forest owner behaviour notwithstanding, as it currently stood, PEFC remained a reactive instrument, established to maintain the status quo. Certification in this instance had become 'a kind of obligatory national forest policy, which everybody gets, and nobody has to do anything'. PEFC was so well established nationally NGOs could have little impact, even if they publicised 'scandalous cases' on the Internet. Business informants preferred to analyse the system in terms of its contribution to what were perceived as 'the well-established international norms of certification'. On this view, the differences between the standards of FSC and PEFC were not particularly great; it was the philosophy underlying the policy processes driving the schemes that was significant. Because PEFC was more 'realistic' than FSC, managers had made commitments to improving their forests; this had freed managers from the burden of legislation, which was now of secondary importance. One interviewee from the third group of informants commented that the advantage in having a system like PEFC was that it could provide a lot of certificates in a short amount of time, but the disadvantage was that there was very little obligation for scheme members to act in accordance with the rules. A second interviewee attributed the lack of change brought about by PEFC to concerns amongst landowners that they would lose money if they pushed the boundaries of SFM. A final interviewee from the third group expressed doubt that forest owners would 'change their habits' under PEFC.

The evidence is persuasive that PEFC was established as a counter-initiative to FSC. However, private forest owners have moved a long way from their original position of outright opposition to certification, to accepting it as a regulatory tool for forest management. The impact of PEFC on behavioural change should therefore be understood as coming from a low base, but the impact of PEFC on shaping the normative behaviour of forest owners should not be underestimated. But it should also be acknowledged that PEFC schemes sit alongside existing codes of forest practice, and therefore simply reflect the status quo of forest management, rather than moving beyond it. These considerations contribute to a rating of **medium**.

Problem-solving

As the PEFC framework is based at the national level it has been argued that it should not be understood as a genuinely international scheme, but as 'a mutual recognition framework through which national certification schemes can recognise each other as having equivalent standards'.[59] The use of the C&I of the MCPFE process and the PEOLG is controversial, since C&I are not designed as the basis for a standards-setting system, nor are they meant to be used to endorse or validate any specific certification scheme. Even with this caveat, the voluntary nature of such guidelines must also be noted. PEFC initiatives are not bound to follow these C&I, and the creation of rules and procedures is left to the discretion of national governing bodies. NGO commentators have been deeply critical of PEFC's refusal to mandate performance-based standards across the system, and argue that: 'There is very little evidence that systems-based schemes alone can lead to environmental improvement'.[60] Where PEFC schemes do have performance-based standards, it is further argued in some instances that they still do not adequately meet social and environmental criteria.[61] The failure of the PEFC framework to address performance issues, it is argued, raises 'serious doubts...about the capability of the system to bring about better forest management at the ground level'.[62]

NGO interviewees were unanimously negative regarding PEFC's problem-solving capacity. One felt that a central question was whether the standard for their country actually required any changes to the existing regime. Simply claiming, as industry did, that it was committed to the standard and that forest owners were aware of and prepared to fulfil its requirements, did not mean the problem was solved. For as long as the clearcutting of ancient and high conservation value forests continued, and was labelled, the problem of biodiversity decline would remain. Most business interests were positive. One argued that a study of PEFC and FSC in Germany had found no differences between the systems at the forest stand level, but that PEFC was stronger in some areas than FSC and vice versa. Another informant, speaking more generally, felt there was little to tell the PEFC and FSC systems apart. However, one interviewee from the third group of informants thought that PEFC was 'a bit below the level' of other standards in their region. It was necessary to recognise that PEFC schemes represented 'small steps in the right direction', and that it was better to have the 'wrong' scheme than not have any. There was extensive debate within all three groups of informants as to whether PEFC's adoption of governmental C&I processes had positively influenced its forest management practices. Views in the NGO group were generally critical, while business interests were positive. Informants from the third group of subjects were mixed. One believed that the C&I processes had brought stakeholders into the system who would otherwise have remained outside, and the C&I endowed the system with a great deal of legitimacy. Another was deeply critical of the way they felt PEFC had manipulated the Helsinki C&I, particularly the PEOLG. They argued PEFC had selectively adopted those elements that fitted their own vision of SFM, and ignored those elements that were costly or difficult for private forest owners to implement. For them, this was a dishonest translation of such processes into the PEFC system.

Attitudes regarding PEFC's problem-solving capacity, particularly regarding environmental performance, are not overly complimentary in either the literature or interviews. Here the problem is not simply whether PEFC's impact on the ground has been effective, but also whether this impact is quantified – especially if performance is not a desired outcome of some standards in the case of process, or systems-based standards. As PEFC was initiated in Europe, where deforestation and other problems such as illegal logging are less of a problem than in developing countries, the impact of the scheme on tackling environmental issues in the tropics has been limited. PEFC's problem-solving ability is rated as **low**.

Durability

PEFC displays a dominance of industry in rule-making. This has resulted in a tightly controlled policy agenda, delivering considerable flexibility in terms of policy options and rule-making – although this can result in lax standards.[63] This has become a problem on account of the lack of any mechanism for ensuring consistency in standards across the system.[64] NGO commentators argue such inconsistencies exist not only across schemes but also within countries.[65]

The problem of inconsistency was a major theme in NGO interviews. In Russia, for example, there were two competitive PEFC systems, one dominated by industry, the other developed by forestry academics. The latter was closely modelled on the FSC, and was identified by one ENGO as having the potential to be beneficial. PEFC standards were also 'highly variable', making some schemes worse than others. From a consumer perspective this meant that when you bought timber or paper with a PEFC logo it did not mean a thing. The advantage to PEFC was that its endorsed programmes were eminently capable of adapting to changing conditions. One business informant acknowledged that there were discrepancies across national systems, but put this down to ongoing processes of continual improvement. National schemes originated from different starting points, and they would 'not all get there at the same time and moment'.

PEFC has changed from being a European to an international certification initiative, and has been substantially revised in the process. This implies a certain ability to adapt to changing market conditions, and to respond to external criticisms. Such flexibility could be viewed positively; alternatively, this constant 'shifting of goal posts' makes determination of PEFC's core values difficult. The lack of consistency across the system means that standards can be either system- or performance-based; certification can be at either a forest management unit, regional or national level; and governance arrangements at the national level also vary considerably. This may reflect either a genuine attempt to respond to local conditions, or expediency. Consideration of

PEFC's adaptability and variability (both positive and negative) gives it a ranking of **medium.**

Conclusions

PEFC's genesis is best understood as a counter-initiative to the market dominance of FSC in Europe, tailored to meet the needs of specific stakeholders, and subsequently expanded to meet the same set of imperatives globally. These origins need not have mattered significantly if they had not had such a profound impact on the structures and processes that underlie the PEFC's governance system as a whole. Internationally, key stakeholders such as environmental NGOs, unions and indigenous peoples are effectively excluded from meaningfully participating in the institution's highest organ, the General Assembly. While they enjoy a degree of representation on the Board of Directors, it does not match the power wielded by forest owners and the forest industry. Although it is technically possible for NGOs or other groups to attend the General Assembly as delegates, or delegation observers, they would attend in a national, not sectoral, capacity. Again, although other international organisations can become non-voting members, none of these represent environmental interests, and are all either forest industry, forest owner, or forestry management oriented. In addition to the low degree of environmental interest representation, the degree of developing country representation is also mixed. This can be partly explained by its early history, but the institution still remains largely Eurocentric in terms of representation in both its Board of Directors and the General Assembly.

PEFC itself acknowledges that its primary participatory framework is at the national and sub-national levels. Nationally, however, the system is ultimately in the hands of forest owners and/or forest industry. Although multi-stakeholders must be invited to participate in standards-setting, there is no such requirement for the earlier formative stages of the national institution itself; in fact it is quite the reverse. Only forest owners and/or the forest industries in a given country can initiate institutional formation. Historically, and constitutionally, the documentation associated with the PEFC makes it quite clear that member schemes are really only accessible to those who support PEFC and its conception of forest management. National bodies have also been accused of a lack of balance in stakeholder representation. In some instances, schemes have been developed entirely by forest owners.

These factors combine to make it structurally difficult for groups in conflict with certain aspects of the scheme to have an effective voice, and may go some way to explaining why certain interests have with-drawn from national schemes. Either unintentionally or intentionally, the system encourages the non-participation of such interests. Non-core stakeholders consequently face an impossible set of choices regarding participation within PEFC. If they do participate, the nature of their relationship to the system affords them only limited influence in com-parison to core interests. If they do not participate, they face the risk of being locked out of the national dialogue around forest manage-ment, since many PEFC schemes comprise, or are fully compatible with, governmental forest programmes. This two-pronged incentive not to participate has led a number of stakeholders to move even closer to the FSC than they might originally have done, thus subjecting them to accusations of bias, making their participation in PEFC even more untenable. This situation has only served to reinforce the divide between environmental–social and industry–owner interests and create pro- and anti-PEFC 'camps'. However, it is as wrong to portray ENGOs in particular as innocent victims or the unwitting subjects of strategic manipulation, as it is to exculpate the PEFC from any blame. But it still leaves matters concerning the social sector, particularly indigenous peoples, unresolved.

PEFC cannot be regarded as a deliberative institution. In terms of decision-making, the very procedures that are supposed to encourage dialogue, such as consensus and qualified majority voting, cannot func-tion effectively because the broader frameworks in which they operate are not functionally democratic. Internationally, there is no universal franchise, and the voting system is weighted in favour of large pro-ducer countries. In view of the NGO–forest owner relations, interaction between parties cannot really be interpreted as being social-political in nature, but merely politicised.

As an institution, PEFC displays some level of competence in terms of implementation. However, due to the many different national govern-ing bodies, country schemes are highly inconsistent, largely on account of different attitudes regarding the adoption of performance- or systems-based standards. Whether PEFC has the ability, will and internal support to implement the far-reaching recommendations of the 2007–2008 gov-ernance review is, at this time, unknown. It is perhaps worth noting here that many of the recommendations in the review are couched in the subjunctive, 'should' rather than the imperative 'shall', and are not mandatory.

6
United Nations Forum on Forests

Historical overview

Origins, institutional development and controversies

The IFF concluded its deliberations in February 2000, and submitted a final report suggesting that forests were in need of a more independent organ, not linked to the United Nations Commission on Sustainable Development (CSD). Negotiations were held to develop a draft resolution, which was submitted to the Economic and Social Council (ECOSOC). This was approved and discussions on the nature of the International Arrangement on Forests began.[1] The various action-related outcomes of the IPF and IFF were codified under a new International Arrangement on Forests, the primary objective of which was to 'promote the management, conservation and sustainable development of all types of forests and to strengthen long-term political commitment to this end'.[2] The six principal functions of this new arrangement were to: a) facilitate the implementation of the IPF/IFF/PfA b) provide a forum for policy development; c) enhance cooperation and coordination amongst relevant agencies and d) enhance cooperation and coordination internationally, through cross-sectoral North/South public–private partnerships at the national, regional and global levels; e) monitor and assess national, regional and global progress on implementation; and finally f) strengthen political commitment (e.g. through ministerial engagement). These objectives were to be facilitated through an intergovernmental body – UNFF. The Forum was given, inter alia, two main tasks: i) within five years, to 'consider...the parameters of a mandate for developing a legal framework on all types of forests' and ii) 'to devise approaches towards appropriate financial and technology transfer support to enable the implementation of sustainable forest management'.[3] These objectives and tasks were to be programmatically implemented by means of the development of a Multi-year Programme of Work (MYPOW).

In September 2000 eight nation-states met with the purpose of working with existing forest-related structures and institutions to develop the concept as well as the basic elements of a new programme of work. The group came to be known as the Eight-Country Initiative and included Australia, Brazil, Canada, France, Iran, Malaysia, Nigeria and Germany. This group provided the background material for an international expert consultation, which produced a non-consensus document canvassing a range of views and ideas on the institution's context and work areas. The Initiative was open to all relevant parties (state and non-state) and was conducted in a transparent manner, and a synthesis report was produced.[4] NGOs were particularly active in the Initiative, and in the light of previous experience, determined to steer the proposed forum in the way they wanted.[5] By the mid-1990s NGOs' views on the value of a forest convention had shifted from an initial position of support immediately post-Rio to outright opposition, based on concerns that a convention would divert attention away from existing initiatives and halt action on the ground while governments negotiated the convention. They were also sceptical regarding the strength of such an instrument, given the historical disagreements between governments and the related refusal to pledge financial resources.[6] NGOs concluded that it would be better to ensure that the new body did not adopt any further proposals, and should instead concentrate on implementing what had already been agreed to. They wanted to concentrate on implementation at the national level, and confine UN-level discussions to reporting and peer review, rather than multilateral negotiation.[7]

A preliminary organisational session of UNFF, held on 12–16 February 2001, adopted two decisions on the location of the UNFF and its future methods of work. The first session (UNFF-1) was held in the UN New York headquarters 11–22 June in the same year, and was mainly administrative, producing draft decisions on the date and venue for UNFF-2, the report of its first session and the provisional agenda for UNFF-2. It adopted three resolutions concerning the the PfA, the content of the MYPOW, and the development of a Collaborative Partnership on Forests (CPF). Provisions were made for the participation of non-state interests referred to as Major Groups at UNFF sessions, via the format of a formal MSD, the first of which would be held at UNFF-2.

UNFF-2, held on 22 June 2001 and from 4 to 15 March 2002, resulted in three resolutions. The first was a ministerial declaration to the World Summit on Sustainable Development, held later the same year. The second related to the implementation of the IPF/IFF PfAs, which had now become the 'plan of action' of UNFF.[8] Countries were to report on the status of their efforts to UNFF-3, and it was determined that the reporting process would be voluntary. The third resolution laid down specific criteria for the creation of a review of the effectiveness of the IAF for presentation at UNFF-5. The six principal functions of the IAF became the reporting criteria for the review, which, it was also determined by member states, should also be voluntary in nature. The UNFF Secretariat and CPF were invited to present a process to facilitate the carrying out of the review at UNFF-4.

The first MSD was held on 6 March 2002. There was general consensus between the countries present and the Major Group representatives that those affected by and implementing forest policies needed to be involved in both planning and decision-making. NGOs in particular emphasised that the Forum 'need not develop more proposals for action, but should instead take concrete actions'.[9]

UNFF-3 was held in Geneva 26 May–6 June 2003. Agreement was reached regarding the composition, terms of reference, scheduling and reporting of three ad hoc expert groups agreed to as part of the intersessional work associated with the MYPOW. Three resolutions were passed regarding the implementation of the PfA and all participants at the session were 'urged...to continue their efforts to implement them'.[10] Two further resolutions were passed regarding enhanced cooperation and policy coordination and strengthening the role of the Secretariat. One other decision of note was a request for guidance from the Secretariat on the format of the voluntary country reports. This had arisen from an identified need for harmony and streamlining to lessen the reporting burden placed upon countries. Two panel discussions were held on the economic aspects of forests and on regional processes and initiatives.

The second MSD was held during UNFF-3 on 27 May. Eight Major Groups participated, two new groups for the first time (women and youth).[11] Discussions regarding the structures and processes associated with forest policy-making at all levels – most notably concerning participation – dominate the summary of the event. Only 3 of the 19 paragraphs do not include matters of governance, and the word participation occurs 12 times in 9 paragraphs. Beyond the broad principle of participation, non-state participants within UNFF were also concerned about a wide range of substantive issues relating to the effectiveness of UNFF as a governance system. These included structural matters regarding interest representation (inclusiveness and resources) and institutional arrangements for openness and transparency, as well as general procedural issues concerning decision-making and implementation.[12]

UNFF-4 was held 3–14 May 2004 in Geneva. Further policy resolutions were adopted regarding the implementation of the IPF/IFF PfA, while a fourth resolution was adopted on the IAF effectiveness review process for UNFF-5, recommending that preparations should be open, transparent and comprehensive in scope to allow for informed decisions at UNFF-5. Once again, it was stressed that the responses from countries, organisations and Major Groups were to be voluntary in nature. It was further agreed that a document would be synthesised from individual reports on the extent of implementation of the IPF/IFF PfA in advance of the fifth session. This would 'provide a global overview of progress towards sustainable forest management . . . as a contribution to discussions' at that session.[13]

One item identified in the MYPOW for consideration at the fourth session, which resulted in no agreement, was the item on traditional forest-related knowledge (TFRK). This item foundered on North/South, as well as internal North/North conflict. The less developed, southern, G77 countries wanted to retain national sovereignty over all commercial decisions relating to TFRK (both access to knowledge and to the benefits derived from such knowledge). They favoured the Convention on Biological Diversity (CBD) as the principal negotiating forum rather than UNFF, as the CBD was developing an agreement on who was to have access to and benefit from the sharing of TFRK already. The developed countries were split between the US and the EU (periodically joined by New Zealand and Canada). The US as a non-signatory to the CBD opposed the idea of an international regime. The conflicts could not be resolved, and the negotiations ended without any agreement.[14]

Two of the ad hoc expert groups, transfer of environmentally sound technologies and finance, and mechanisms for monitoring, assessment and reporting, were scheduled to present their reports at UNFF-4. Negotiations inside the former resulted in disputes over funding via official development assistance (ODA), favoured by the South, versus private sector financing preferred by, and economically advantageous to, the North. A relatively weak decision (rather than a resolution) was passed encouraging 'further consideration' through the MYPOW.[15] The second group – following what had by now become the norm in UNFF – recommended that monitoring, reporting and assessment, as well as adoption or implementation of C&I should be 'on a voluntary basis in accordance with national priorities and conditions', and would not include NGO input.[16] This led the writers of the *Earth Negotiations Bulletin* to conclude that the lack of input from civil society would exacerbate their feelings of alienation 'and could eventually deprive the post-UNFF arrangement of an important source of legitimacy'.[17]

The third MSD was held on 6 May 2004. The dialogue was in two parts, with the first component consisting of general policy discussion arising from the Forum themes of TFRK and the social and cultural aspects of forests. The second component was more focused on the implementation-related topics of capacity-building and partnerships.[18] The dialogue regarding TFRK reflected the events occurring within the Forum's own negotiations and repeated some of the unresolved NGO frustrations of IFF/IPF.[19] By the end of UNFF-4, the series of failed negotiations between governments resulted in a sense of frustration amongst non-state participants. The last statement delivered at the plenary by Indigenous Peoples' Organisations (IPOs) and NGOs highlighted their concerns regarding both the outcomes and processes of the session.[20] The MSD itself was criticised by these groups as being 'just an exercise in window dressing... the outcomes and concerns expressed during these discussions never found their way into the important decision-making and text negotiations carried out by the governments'.[21] The *Earth Negotiations Bulletin* was equally frank: 'One point of clear consensus in Geneva was that UNFF has failed to deliver on its stated aims, and that continuing the arrangement in its current form is neither politically viable nor desirable.'[22]

UNFF-5 was held in New York 14 May 2004 and 16–27 May 2005. Only one draft resolution was agreed, regarding the report of the fifth session and a provisional agenda for UNFF-6, referring the matters to ECOSOC. Two decisions were made, one regarding the accreditation of intergovernmental organisations, the other concerning the IAF review process. With respect to the latter it was decided to complete the consideration at UNFF-6, which was to be guided by draft text developed in the course of informal consultations during UNFF-5. UNFF-6 was also given the task of creating an agenda for UNFF-7, and the Forum's life, by implication, was further extended. Two round-table discussions were held during the course of the high-level, ministerial, segment of the session, on restoring the world's forests and forest law and governance. The second addressed the problem of illegal logging identified during earlier IPF discussions, and noted that initiatives external to UNFF, including certification and public procurement policies, were valuable tools for providing market access for legal and sustainably managed forest products.

Much of the Forum was occupied by discussions concerning the report prepared by the third (and extremely long-titled) Ad Hoc Expert Group on Consideration With a View to Recommending the Parameters of a Mandate for Developing a Legal Framework on All Types of Forests. The title of the group was deliberately equivocal, and reflected the ongoing debate over the type of forest instrument UNFF should adopt. The report presented two basic options, built around either a non-legally binding instrument (NLBI), or a LBI, both of which deeply affected the future shape and direction of the IAF. The report was hotly debated, split between the US and the G77 (in favour of the NLBI route) and the EU (pro-LBI). The debate as to whether the IAF should be legally binding or not dated back to earlier disagreements. Neither the UNCED nor IPF/IFF had ever fully resolved the debate regarding a forest convention. A second element of the conflict at UNFF-5, again repeating old arguments, related to funding the implementation of the IAF. The G77 pressed its claim for funding to implement the IAF, while the EU wanted clear targets and timetables. In the end, the text generated was so weak that there was nothing to present at the scheduled ministerial declaration.

Another major source of conflict, this time between countries and non-state interests, concerned the Secretariat's attempt to integrate the MSD into the high-level ministerial discussions during the session. This was strongly opposed by Cuba, and the matter almost went to a vote. It was finally agreed by member states that from UNFF-6 onwards Major Groups would give their statements only after governmental delegations had presented. The impact on Major Groups was that they were essentially relegated to the role of passive observers. This was condemned by NGOs as a 'retrograde position', which violated the spirit of Agenda 21.[23] The fourth MSD was held on 25 May. Representatives of the Major Groups simply read a series of prepared statements and there was no discussion.[24] Business and industry interests were equivocal regarding the IAF, neither supporting nor opposing any specific type of international arrangement, but asking for greater private sector and non-governmental participation. They recommended a set of basic principles and minimum requirements for greater coordination of forest policies and paying more recognition to the trade in sustainably managed forest products.[25] NGOs asserted that ten years' debate regarding whether there should be a global forest convention had prevented progress on a range of issues and that UNFF needed reform.[26] Criticising

Major Groups' participation in the MSDs 'as a way to segregate the input provided by those stakeholders' they argued that without 'radical changes to ensure the effective consideration of proposals' there was no point in continuing with them.[27] The Forum ended with the chair, and a range of countries, expressing disappointment that UNFF had not risen to the challenge of the forest crisis, and was now in danger of becoming increasingly peripheral to the international dialogue.

UNFF-6 was held in New York 27 May 2005 and 13–24 February 2006. There were no resolutions and only one decision passed (relating to the accreditation of intergovernmental organisations). However, a draft resolution for adoption by ECOSOC was agreed to, representing some progress in deliberations between the member states. Firstly, and as a consequence of the review of the effectiveness of IAF, it was agreed that the IAF would contain three new principal functions in addition to those referred to in ECOSOC resolution 2000/35: enhancing UNFF's contribution to existing international agreements, such as the Millennium Development Goals (MDG); encouraging countries with low forest cover to increase conservation and rehabilitation activities; and strengthening interaction between UNFF and other regional and sub-regional forest-related entities. Secondly, four new global objectives on forests (GOF) were identified as means of implementing the IAF, summarised as follows: 1) reverse the loss of forest cover and increase efforts to prevent forest degradation; 2) enhance forest benefits and their contribution to international development goals; 3) increase the area of protected forests and areas of sustainably managed forests; and 4) reverse the decline in ODA for SFM. It was also agreed that following its seventh session UNFF would meet biennially on the basis of a more focused MYPOW to be adopted by the Forum at its seventh session.[28] Regional meetings and country-, region- and organisation-led initiatives (CLI, RLI, OLI) would occur in between, concentrating on matters implementation and the 2007–2015 MYPOW.[29] After a decade and a half of stalled negotiations on a LBI, all parties agreed to the adoption of a NLBI at UNFF-7. On that basis, the effectiveness of the IAF would now be reviewed in 2015, and a full range of options, (including an LBI, which its supporters managed to keep on the agenda), would be considered then. It was also agreed that following its seventh session UNFF would meet biennially, with regional meetings in between, concentrating on implementation.

Owing to the interventions of Cuba and other countries at UNFF-5, the status of the fifth MSD was much reduced, producing the lowest level of non-state engagement since talks regarding the IAF began. The lack of engagement by Major Groups led the writers of the *Earth Negotiations Bulletin* to conclude that NGOs were better off working through other agreements such as the CSD and/or the Forest Stewardship Council (FSC), where they had 'better luck pursuing their agenda'.[30]

Ironically, given the new arrangements for UNFF sessions, Major Group interests participated actively in the subsequent meeting of the expert group on the NLBI, held in New York, 11–15 December 2006. Several of their interventions were included along with the rest of the bracketed text produced by country participants.[31] This may reflect the changed circumstances arising from the decision to postpone discussions regarding a forest convention, and with previous divisions between hard and soft law becoming increasingly blurred, the NLBI discussions occurred somewhere on that continuum. A consensus vision document and the support NGOs received from various countries for their ongoing involvement may have strengthened their resolve to stay involved at some level.[32] It has also been argued that the increasing proliferation of various soft law approaches to forest management regulation, exemplified by the forest law, enforcement and governance (FLEG) processes (see below), further moved the debate beyond a forest convention, which was not supported by NGOs.[33]

UNFF-7 was held on 24 February 2006 and 16–27 April 2007 in New York. One resolution, concerning the MYPOW for the period 2007–2015, was passed, and one draft resolution, concerning the NLBI, was referred to ECOSOC for adoption. Three draft decisions were also referred to the Council relating to the UNFF Bureau's term of office, the dates and venue for UNFF-8, the report of UNFF-7 and agenda for UNFF-8. MYPOW discussions centred upon operationalising UNFF's activities for UNFF-8, -9, -10 and -11. It was agreed these activities would be based around the GOF identified at UNFF-6, the implementation of the NLBI, and examining the IAF (to occur at UNFF-11). Various delegates expressed views that the MYPOW represented the 'true value-added' agreement on UNFF-7, managing to avoid difficult questions regarding funding and the NLBI, and producing a 'more ambitious' programme than that originally agreed to at UNFF-1.[34] The success of these discussions led the writers of the *Earth Negotiations Bulletin* to conclude that this 'could be the start of a transition of UNFF's function from being a forum burdened with an intricate negotiating task into an institution that generates useful information and facilitates cooperation'.[35]

However, whilst progress on the MYPOW was positive, other delegates felt that the NLBI was by contrast 'a distracting side-show'.[36]

Afraid that UNFF would collapse without agreement on the NLBI, its production in the last hours of the meeting was an achievement in itself, even if there were deep concerns that the language in the final text weakened much of what had already been agreed to in previous documents. A number of key elements, discussed at length during the meeting, including a process to facilitate NLBI implementation, the use of time-bound and measurable targets, combating illegal logging, and a definition of SFM were all dropped.[37] The language relating to funding of the NLBI was vague and left for future negotiations. This resulted in a somewhat 'meatless' NLBI, which served to reinforce the views of some – mostly European – countries that they were better off developing proposals outside UNFF, and a closed meeting was held to do just that.[38]

Two MSDs were held 18 and 23 April. According to one participant, the general rate of NGO participation was 'very low' and many previously active environmental NGOs were entirely absent.[39] Nevertheless, a number of governments expressed their support for Major Group involvement in UNFF, and a proposal for a non-state initiative to be hosted by a member state of the Forum was hailed as 'a good sign that stakeholder participation was evolving from dialogue to concrete action on the ground'.[40]

UNFF-8 was held 27 April 2007 and 20 April–1 May 2009 in New York. One resolution was passed, concerning enhanced cooperation and coordination. It was also decided that agenda item six, 'means of implementation for sustainable forest management', should be held over until UNFF-9. This was a somewhat ominous development, since this agenda item cut to the heart of the new NLBI, and a great deal of preparatory discussions had already taken place by means of the new – also extraordinarily long-titled – Open-ended Ad Hoc Expert Group to Develop Proposals for the Development of a Voluntary Global Financial Mechanism/Portfolio Approach/Forest Financing Framework. Some familiar patterns re-emerged over the course of the session. Once again, national reporting on implementation (in this case of the GOFs) was low. Member states also defaulted to their traditionally polarised camps on matters of financing: China and the G-77 argued for the establishment of a global forest fund, while the donor countries argued that the private sector and other investment mechanisms should be used to encourage SFM. Despite all-night meetings and ongoing daytime discussions, no compromise text eventuated. The lack of agreement on funding the NLBI effectively stalled the positive progress made at UNFF-7. According to *Earth Negotiations Bulletin*, UNFF had once again demonstrated that it was ultimately 'maladjusted to stimulating interactive dialogue', and had fallen victim to ongoing conflict around 'forbidden' international forest policy topics, including tenure rights, illegal logging and forest certification.[41]

Two MSDs were held on 22 and 28 April. NGOs noted the increasing policy dominance of the climate debate on forest discussions and concerns were expressed about the emerging UN Programme on Reducing Emissions from Deforestation and Forest Degradation in Developing Countries (REDD), which, it was felt, could have 'significant impacts on the rights and governance structures' of indigenous peoples and local communities.[42] Interestingly, UNFF Director Jan McAlpine exhorted Major Groups to improve their level of engagement with member states. This was, however, a somewhat tall order, given their restricted role, which post UNFF-5 had effectively confined them to making general statements 'leaving many delegates less than fully engaged, if present at all'.[43]

Governance within UNFF

System participants

There are three broad constituencies in UNFF: member states, intergovernmental agencies working on forests, and the so-called Major Groups referred to in *Agenda 21*. Member state representation in the Forum itself differs from the previous IPF/IFF arrangements under CSD. Although both CSD and UNFF are subsidiary organs of ECOSOC, UNFF has a higher status in the sense that it consists of a universal membership (all UN member states are members of UNFF) whilst CSD has limited membership. Governmental participation reflects this profile, with negotiations usually conducted by diplomats and high-level national delegations. There are also ministerial segments conducted at important sessions.

Resolution 1/1 of UNFF-1 reiterated the importance of stakeholder participation and instituted the concept of an MSD at each session to engage representatives of five key (non-state) Major Group stakeholders. These stakeholders are identified as: forest-related NGOs (e.g. Greenpeace International); indigenous people (e.g. the Forest Peoples Programme); scientific and technological communities working in forest-related fields (e.g. IUFRO); business and industry related to forests (e.g. WBCSD); and forest owners (e.g. Confederation of European Private Forest Owners – CEPF). To this should be added women, children and youth, local authorities and farmers, which are identified in *Agenda 21*, and recognised in subsequent UNFF literature, making a total of nine identified Major Groups.

Institutional arrangements

International level

A Bureau from within the Forum was also created at UNFF's inception and consists of one chairperson and four vice-chairpersons, one of whom acts as rapporteur, with reports being submitted to ECOSOC and through it to the UN General Assembly. A Secretariat was established through the UN Secretary General, consisting of its own, and other, occasional, staff from the various other secretariats and agencies. The role of the Secretariat is to serve the Forum, coordinate its activities with the Secretariat of the CBD and support the CPF. The CPF was created in 2001 to enhance coordination and collaboration between agencies working on forest-related activities. It is constituted from the various forest-related international organisations, secretariats of related conventions and institutions working on forests to support the work of UNFF. In addition to plenaries and closed discussions, UNFF sessions also include panel discussions, led by experts who present information on specific topic areas and field questions from the audience. Ad hoc expert group meetings, where experts 'deliberate and provide advice on scientific and technical issues related to forests, as well as advancing the objectives of UNFF', also take place between UNFF sessions.[44] Country- and CPF-led inter-sessional initiatives have also been encouraged by UNFF. These are for the purpose of 'catalysing enhanced cooperation and coordination where complex and politically sensitive issues are discussed and analysed and tabled at UNFF sessions for further deliberations and decisions'.[45] See Figure 6.1 below.

National level

UNFF forest management-related objectives are implemented through NFPs (originally by means of the PfA).[46] NFPs have been described as 'policy planning instruments, striving to render politics more rational, more long-term oriented, and better coordinated by a series of *basic principles and elements* that replace the principles of traditional technocratic planning'.[47] The development of NFPs are assisted at the international level by two agencies outside the UNFF: the FAO's NFP Facility, in operation since 2002 and an offshoot of the Tropical Forestry Action

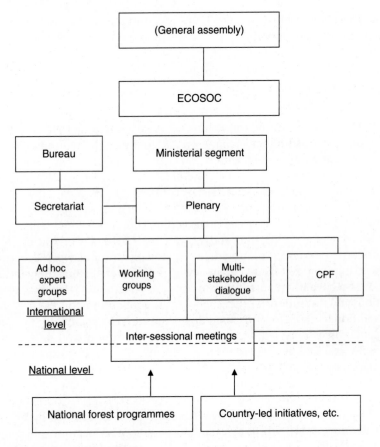

Figure 6.1 Structure of UNFF.

Plan (TFAP); and the World Bank's Programme on Forests (PROFOR), initially created by the United Nations Development Programme (UNDP) in 1997 and moved to the World Bank in 2002.

Standards development

UNFF does not develop standards. Its activities are guided instead by the IAF. Among IAF's most important elements, carried over from the IPF/IFF processes, were the PfA. UNFF's contribution was to streamline the 270 PfA into thematic elements, which constituted specific agenda items at UNFF's second, third and fourth sessions.[48] Post UNFF-6, however, the orientation of the IAF was changed somewhat, with greater emphasis now placed on the GOF, which are described as providing 'clear guidance on the future work of the Forum'.[49] The thematic components of the PfA have not been completely abandoned, but according to one participant it is now 'up to countries whether to use all or some of them for national reporting'.[50]

Institutional typology

Member countries within the UNFF are by far the most influential actors, reflecting their status in the UN system as a whole. States negotiate text, pass resolutions and make decisions, and their power is absolute. UNFF can, however, be influenced by non-state interests through the Major Groups that participate in the MSD, by intergovernmental organisations via the CPF, and through the role played by special agencies (as well as state entities) in implementing SFM on the ground via NFPs and other mechanisms. Non-state interests sometimes also attend meetings as part of state delegations, but they do not sit in on closed sessions. These qualifications provide a moderating influence, and although UNFF clearly belongs on the state-centric end of the authority continuum, it is rated as being between **high** and **medium**.

UNFF conforms largely to the aggregative model of democracy on account of its majority voting system, but conformity is not complete, since by tradition voting powers remain unexercised in favour of full consensus agreement (see below). Nevertheless, nation-states themselves are aggregated around various veto coalitions and voting blocs that represent specific – competing – interest groupings akin to political parties (e.g. those supporting either an LBI or an NLBI). Although UNFF therefore sits on the aggregative end of the democracy continuum, formal voting is not practiced, and much of the interaction between enfranchised participants is discursive in nature, earning it a rating of **low.**

Although largely steered by the state and top-down in nature, UNFF nevertheless permits a degree of interaction with non-state and national-level interests. In terms of its structure (particularly with regard to the subordination of the Ad Hoc Working Groups to Plenary, and Plenary in turn to the Ministerial Segment, and thence to ECOSOC and the UN General Assembly), it is best described as a form of hierarchical governance, albeit with some 'open' or 'mixed mode' tendencies, given the social–political nature of its mandate.[51]

From its foundation UNFF decided to foster international cooperation through public–private partnerships (PPPs).[52] It is consequently associated with, but not directly responsible for, a range of 'new' governance arrangements at national and regional, as well as international,

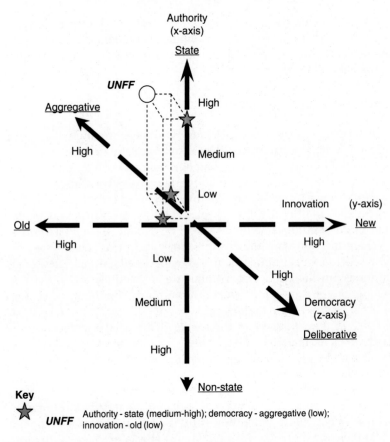

Figure 6.2 Institutional classification of UNFF.

levels. These are noted as being instrumental in promoting SFM and for addressing issues regarding illegal logging and good governance.[53] It is best placed somewhere in the middle of the governance continuum; given its top-down structure, it is more directly associated with 'old' governance arrangements than 'new', but with a rating of **low**. See Figure 6.2 above.

Governance quality of UNFF

Commentary

UNFF received 24 points out of a maximum total of 55 (See Table 6.1 below). Nine indicators achieved low ratings (equality, resources,

Table 6.1 Evaluative matrix of UNFF governance quality

1. Meaningful Participation

Principle	1. Meaningful Participation		
Criterion	1. Interest representation Highest possible score: 15 Lowest possible score: 3 Actual score: 7	2. Organisational responsibility Highest possible score: 10 Lowest possible score: 2 Actual score: 4	Sub-total (out of 25): 11
Indicator	Inclusiveness	Equality	Resources	Accountability	Transparency

	Inclusiveness	Equality	Resources	Accountability	Transparency
Very high					
High					
Medium	3				
Low		2	2	2	2
Very low					

2. Productive deliberation

Principle	2. Productive deliberation		
Criterion	3. Decision-making Highest possible score: 15 Lowest possible score: 3 Actual score: 6	4. Implementation Highest possible score: 15 Lowest possible score: 3 Actual score: 7	Sub-total (out of 30): 13
Indicator	Democracy	Agreement	Dispute settlement	Behaviour change	Problem solving	Durability

	Democracy	Agreement	Dispute settlement	Behaviour change	Problem solving	Durability
Very high						
High						
Medium						3
Low	2	2	2	2	2	
Very low						

Total (Out of 55) 24

accountability, transparency, democracy, agreement, dispute settlement, behaviour change and problem-solving), and two medium scores (inclusiveness and durability). There were no very low results. The conventional pass/fail target value of 50 per cent was not met by any criterion, the results being 47 per cent (interest representation and implementation) and 40 per cent (organisational responsibility and decision-making). At the principle level, in terms of meaningful participation the aggregate result was 40 per cent and for productive deliberation, 43 per cent.

Evaluation

Interest representation

Inclusiveness

UNFF is the only subsidiary body of ECOSOC with universal country membership, which has given it a greatly enhanced profile within the UN system. Non-state participation is more restricted. Resolution 1/1 outlines the rules of procedure under which Major Groups can participate in UNFF sessions. These were based on the rules governing functional commissions of ECOSOC, and the supplementary arrangements established by ECOSOC for the CBD. The Forum was expected to extend the 'participatory practices' established by the CBD, IPF and IFF.[54] Under these procedures participating non-state interests are permitted to attend as observers. They may also, at the discretion of the chair, make oral interventions, with the consent of members, and may also be consulted or heard by any relevant committees.

The output of the involvement of the Major Groups in the MSD is a Chairman's Summary, included in the final report of each Forum. Any proposals arising from the segment are presented to the Forum and 'taken into careful consideration' in any negotiated decision; this is considered an 'efficient and effective way to involve Major Groups'.[55] UNFF considers the Major Groups to be 'actively involved in UNFF and its programmes'.[56] This role, however, was undermined by the changes to the MSD instituted as a result of Cuba's intervention at UNFF-5. It has been argued that Cuba initially attempted to prevent the inclusion of the MSD at UNFF-6 all together. The final decision to allow Major Groups to speak only after government delegations represented a compromise, but one which, it has been alleged, nevertheless turned the discussions of the MSD at UNFF-6, into 'informal side events... rather than as part of the formal proceedings. Not surprisingly, the result of this decision was significantly lower Major Group involvement.'[57] The MSD is criticised for failing to live up to its promise, creating a 'participatory ghetto' for Major Groups.[58] This has resulted in the waning of interest and participation of Major Groups, particularly in the wake of UNFF-5, and the objection to Major Group involvement expressed by some governments. Although UNFF claims to have taken action to increase participation at various levels it has accepted that 'further efforts are now needed'.[59] Commentators have called for UNFF 'to increase the opportunities for meaningful participation by multiple stakeholders'.[60] The UNFF is also criticised as being out of step with more inclusive policy processes, and the loss of NGO participation is attributed directly to this shortcoming. In contrast, such processes as the FLEG and the Asia Forest Partnership (AFP) have made efforts to include a broader range of actors and stakeholder perspectives.

A wide range of comments was made concerning the inclusiveness of UNFF. The view was expressed across informants from group three that for the nation-states, the inclusive nature of UNFF was a good thing. One of the strengths of UNFF was that it had universal membership, which was not usual in the UN system, particularly as a subsidiary organ of the ECOSOC, which did not usually encourage such arrangements. The presence of in excess of 180 countries meeting and engaging around discussions about illegal logging, enabling instruments, work programmes, and the goals and objectives of UNFF, was empowering. NGOs were generally negative. Their involvement could not be counted as participation because they could not make any changes to text. After efforts to exclude NGOs their status had been reduced to observers only in plenary sessions. After UNFF-6 it had become even more difficult to convince NGOs to stay involved. Participation in UNFF could only be described as lobbying. The only value it had was that it provided opportunities to meet and discuss broader forest issues with colleagues. One comment sums up NGO feelings and is worth reproducing in full:

> I was always really depressed after the UNFF. You work really, really hard for two weeks, and in the end there's nothing to show for it. You're just empty. The NGOs usually walk out. That's not a nice thing to do, but it's just that you've been running after all these people, and you've tried to make your statements, or give delegates a statement in the hope that they would read it or just discuss it, but they are far too busy to listen to you, let alone read it. It's not a nice process at all.

Interviewees from group three of informants were mostly sympathetic towards the inclusion of non-state interests. The reduced Major Group participation brought about post UNFF-5 meant that a substantive part of the forest agenda was missing, which was a problem. The MSD was far too restricted to be of any use. Most government delegates took it as an opportunity to go shopping. They generally wanted to see more, not less, involvement but noted there is an underlying tension between member states in the UNFF system over the level of non-state inclusion. One interviewee attributed the push by some governments to reduce participation to the inclusion of NGOs in some delegations. Some governments wanted to be able to negotiate in an environment that was free of NGO influence.

For nation-state participants, UNFF is highly inclusive and all sub-components of the system are open to their input. In terms of non-state participants, there is also a relatively broad range of interests represented within the Major Groups, although some commentators call for a greater spread of stakeholders. Non-state interests do not participate as actively as member states since they generally only have observer status in most structures, and a restricted degree of involvement in others, with the notable exception of the ad hoc group for the NLBI at UNFF-6. Although state interests experience a high degree of inclusiveness, non-state interests do not. These high and low degrees of inclusiveness are counterbalanced, leaving the system as a whole somewhere in the middle between inclusive and exclusive, with a rating of **medium**.

Equality

Scholars recognise that the different forms of new forest governance need to be reconciled with contemporary demands for participation and the protection of rights. They see the necessity of having a general frame and purpose for forest governance that can be identified as legitimate by all concerned. Such a frame should address what they consider to be the most important and demanding linkage between all levels of SFM: the problem of social justice.[61] They envisage justice in terms of economic equality, whilst other observers relate justice and equity to how forest laws are framed and implemented, emphasising the linkages between good governance, participation and sustainability. Meaningful participation is seen as being essential for SFM. UNFF is accused of unnecessarily constraining and even denigrating civil society participation and there have been calls for it to embrace non-state interests as equal partners. Other processes such as the CBD have been identified by NGOs as providing much greater access. This may go some way to explaining the growing stridency in the criticisms made by some non-state interests over the course of UNFF. As early as UNFF-2 NGOs called for UNFF-related entities to ensure greater balance of interests in its activities at international, national and local levels, and singled out the CPF in this regard.

Perspectives on the nature, degree and expression of equality varied between non-state and state interests. According to the latter group, UNFF was a government-led process under the framework of the UN, and this was why government delegates had better access. Non-state involvement was problematic because a number of countries in the process had very strong feelings either for or against the engagement and participation of the Major Groups. Some countries had threatened to walk out of discussions on account of what role Major Groups should or should not be allowed to play. This explained why the MSD was so 'completely ineffective', and why so much of the activity actually took place in the corridors between meetings. On the non-state front, one NGO informant described the nature of their participation at UNFF in comparison to IPF/IFF, where they had actually been involved in drafting some of the original PfA, including those on TFRK and the underlying causes of deforestation. UNFF had proved to offer little more than an opportunity to lobby governments, such as those in the EU, usually resulting in little action, which was 'very disappointing'. Another interviewee from the business sector felt that a number of governments, as well as the Secretariat, were not really interested in multi-stakeholder input, and was deeply critical of the lack of status accorded to the MSD.

Nation-states have extensive opportunity and the capacity to participate in UNFF's structures of governance. They can vote, negotiate text, draft decisions and make agreements. Non-state interests cannot. These rights of access give nation-states a much greater capacity to participate in, and make use of, UNFF's facilities than non-state interests, since the latter's access is more restricted. Nation-states have a high degree of influence within the UNFF system and over its outcomes, dependent on their geo-political power and strategic alliances. Non-state interests are less able to influence the UNFF system since their involvement is relatively restricted under ECOSOC rules. Their power is based largely upon their ability to lobby effectively, and they cannot directly affect the outcomes of sessions. Nevertheless, the UNFF is characterised by a low degree of equality amongst its participants, with a rating of **low**.

Resources

Forest-related programmes can be both time and resource intensive, and the processes themselves require adequate resources for facilitation and management. UNFF's effectiveness at this level has been challenged, since NGOs identified financial constraints in the MSD of UNFF-4 as the 'main obstacle' to national and local capacity-building and implementation.[62] UNFF itself has recognised that capacity-building and technical and financial assistance are 'crucial preconditions for strengthening governance systems'.[63] Institutionally, it has also acknowledged the need to assist stakeholder groups that have limited capacity or opportunity to participate and at UNFF-3 it was agreed that the Forum should formally constitute a Trust Fund, which should be used for the assistance of developing country delegates to participate in future meetings. By UNFF-6, the Bureau identified an 'overwhelming' need for travel support from participating developing countries, which exceeded the allocated budget of approximately $120,000 by some $350,000, and additional funds had to be sourced from donor countries.[64] Some commentators consider UNFF to be 'ill-equipped' as an international forest governance system, largely as a result of resource constraints.[65] Regrettably, it has been unable to obtain resources for SFM by reversing the decline in ODA from donor nations. This trend is likely to continue 'unless it addresses the fundamental structural and procedural weakness that currently limit its effectiveness'.[66]

The comments from interviewees regarding the resource-related aspects of participation and interest representation in UNFF are thematic, rather than partisan, in nature. The cost of attendance is a cross-sectoral concern. Distance from New York and the cost of accommodation were identified as barriers to participation by several attendees. Others noted that UNFF was costly in terms of the amount of time it took up, as well as being financially draining. The time spent in comparison to the generally unsatisfactory outcomes that were negotiated led several to agree with NGO criticisms that the money could be better spent. Holding regional workshops, for example, was more effective. The provision of funds by UNFF itself, and by governments for national interests to attend was also discussed. UNFF received a regular budget from the UN and some money for travel support for developing countries; there was usually at least one delegate from each developing country, about 50 all told. Some funds were made available to nongovernmental representatives from 2004 onwards. Generally, the UNFF Secretariat preferred to encourage governments to include non-state interests in their delegations, and to find other organisations to support that, such as FAO. Another issue, and one that led to expressions of frustration amongst some interviewees, was that there was no real financing mechanism for the UNFF itself. The EU and a number of individual EU member countries had consequently invested considerable amounts of money, but there were no broader funding mechanisms for implementation, leading some member states to seriously consider abandoning the process.

UNFF has put some effort into securing funds for developing country participation. The establishment of a facility within the Trust Fund for developing countries, as well as the provision of resources from donor nations, are valuable contributions. The situation for the Major Groups is less satisfactory. Although some funds are available for NGOs, these are only on a quasi-official level, and still insufficient in terms of providing for widespread representation. This is commendable, but on a broader level there is a severe lack of funds to support the administration of the system. There are also ongoing conflicts over funding means of implementation. All these factors combine to produce a rating of **low**.

Organisational responsibility

Accountability

In the broader UNFF system, NFPs, model forest networks and community-based forestry are the principal models of forest partnership on the national, global and local levels. Governance scholars have noted the inherent problems of accountability in network governance and the democratic challenges associated PPPs pose. During the MSD held at UNFF-4 NGOs stressed that effective partnerships between governments and non-state interests required clear agreements and jointly defined roles and responsibilities, reflecting the will of participating constituencies. Accountability in such relations was identified as being essential in ensuring the credibility of such approaches. NGOs expressed a note of deep concern that some types of partnership were 'often established without the adequate participation of civil society'.[67] UNFF's ad hoc expert groups consist of experts designated by governments but nominated in their personal capacities, not as governmental delegates. The intention behind this is to avoid promotion of government positions. However, this arrangement has led one commentator to observe that 'many governments appointed senior negotiators to the expert groups, with the unsurprising result that the views expressed by experts often bore a striking resemblance to government policy'.[68]

All three groups commented on the degree to which government delegates could really be considered accountable to other participants in the UNFF system. According to one interviewee, one of the biggest problems with UNFF was that governments, who were also the largest forest owners, did all the negotiating. They would not stand for a process that criticised how they managed their forests in any way. Corruption was also a problem with some member states. Money given to Russia, or many African countries, or Indonesia, never got to local communities, and either went to the banks in their capital cities, or was simply stolen and sent to Switzerland or Luxembourg. One governmental interviewee commented that it was essential that the money their country invested had a sustainable impact on the ground. They needed to demonstrate to their parliament and taxpayers that their contributions had achieved something to reduce the 15 million hectares of global forest loss per year. They wanted clear rules and verification obligations for countries giving, and in receipt of, funds, as these could be manipulated for strategic political purposes. The US was a very large donor and had created an 'unholy alliance' with Brazil to block the proposed LBI. At least half a dozen members of the EU were also major donors but they wanted the LBI. Despite the aid Brazil received from EU members, the economic forces backing its rejection of any kind of global forestry instrument were stronger. Further comments were made by other interviewees about the behind-the-scenes activities of the UNFF Secretariat prior to sessions, and the lobbying they were subjected to by member states; there was no imputation of misconduct by the Secretariat in this regard, but the potential was there. The Secretariat also at times ran its own agenda. Some issues were discussed at Forums that were more about creating a higher profile for the Secretariat within the UN system than meeting the needs of member states.

UNFF, like other intergovernmental processes, operates beyond the mandate of national-level parliaments, and its degree of 'vertical' accountability in terms of the direct responsibility of national governments to their domestic constituents is questionable. However, it should be noted that the existence of the linkage back to national-level forest processes via NFPs partially mitigates this. Nevertheless, there is a compelling case to be made that there is little opportunity for member states to be held accountable within the UNFF framework for their forest-related activities, particularly since all the main intergovernmental instruments are voluntary. Within the Forum itself it could be argued that there is also little 'horizontal' accountability of member states to other participating interests. Further accountability problems exist in relationship to the role played by government appointees in the various ad hoc committees, and there is some ambiguity regarding the role played by the Secretariat. Monies provided by donor countries can at times be highly politicised, and used to 'buy' favours. Accountability in the UNFF system is **low**.

Transparency

UNFF has acknowledged transparency as an 'indispensable attribute of good governance'.[69] Institutionally, it was also committed from its inception to build on the 'transparent...practices' established by the CBD, IPF and IFF.[70] Documentation also comments on the importance of 'transparent and participatory practices, including multi-stakeholder participation at the national level' as part of the requirements for implementing SFM 'in a cohesive and comprehensive manner'.[71] The reporting on implementation of the IAF is also expected to operate in a transparent manner. In order to meet this commitment, it was agreed at UNFF-2 that country reports, as part of the review process, would be made available at UNFF-3 and subsequently.[72] In the period after UNFF-4 and prior to UNFF-5, participants were requested to report on their forest-related activities, for consideration at UNFF-5. The study was in two parts: an analysis of the progress in implementing the IPF/IFF PfA, according to the 16 thematic elements, and a review of the effectiveness of the IAF according to six principal functions.[73] However, it was determined that the reporting process should be voluntary (hence the term Voluntary National Reports).[74] Director Jan McAlpine 'lamented' the low level of country reporting at UNFF-8.[75] Twenty countries provided national reports (compared with 14 for UNFF-2, 39 for UNFF-3, 38 for UNFF-4 and 55 for UNFF-5).

Discussion amongst interviewees relating to transparency within UNFF focused on comments about the Forum itself and the role played by governments. One NGO informant criticised the failure of UNFF to insist on strong transparency mechanisms for national reporting. It had developed a weak process, oriented around a questionnaire with few substantive reporting requirements, and which few countries had filled in. With such inadequate reporting it was impossible to tell what was really being done. Very few countries reported on their activities more than once, and a large number had not reported on their activities at all. A number of informants from group three were critical of government behaviour. The way in which governments met to negotiate text was not open. It was true that UNFF was an intergovernmental body, and that governments were its major clients, but in terms of transparency, the opportunities for either the members of the CPF, the Major Groups or any other NGO stakeholders to participate were 'very limited'. Another contrasted UNFF's lack of transparency to the MCPFE in Europe, where draft resolutions were still negotiated by the member states, but all interests were present during these negotiations. In UNFF all major negotiations occurred behind 'closed doors'. This interviewee wondered whether the outcomes and the generally constructive nature of discussions in the MCPFE were a consequence of the greater transparency of the process.

There are at least four issues relating to an evaluation of the transparency of UNFF. The first relates to the lack of regular – and consistent – national-level reporting. Secondly, relating to openness, is the manner in which member states engage in negotiations (usually in sessions generally closed to non-state interests). Thirdly, concerning the availability of, and access to, information, UNFF's performance is mixed. Its past, present and proposed future activities are freely available on the Internet, but the session reports are couched in largely neutral terms, and do not discuss the diplomatic nuances of any deliberations (this is left to such external publications the *Earth Negotiations Bulletin*). Finally, detailed financial reporting, such as income and expenditure, the allocation and amount of travel expenses and *per diems*, and donor country contributions to the Trust Fund is not readily available.[76] Transparency in UNFF is **low**.

Decision-making

Democracy

UNFF follows ECOSOC's *Rules of Procedure* whereby each member state has one vote. By convention UN and its processes generally function via consensus (see under agreement below). Non-state interests do not have the right to vote. Any rights they are given depend on the type of organisation, although the categories are not entirely clear. Specialised agencies cannot vote, but may be represented at meetings and participate in deliberations that relate to items of concern to them, and may submit proposals regarding such items. Other intergovernmental organisations accorded permanent observer status by the General Assembly may participate in deliberations of relevance to their activities without the right to vote. NGOs may be granted consultative status according to the determination of the Committee on NGOs.[77] These rules were supplemented by arrangements that permitted NGOs to make written representations and speak (with permission) but not have any negotiating role.[78] These slight amendments, which granted NGOs the right to speak on the general business of UNFF, were not, it was stressed, to be seen as setting a precedent.[79]

The impression created by interview subjects was that the degree of democracy experienced at UNFF depended on the extent of participation in the system. According to one actively participating informant, because the UN was a body made up of member states, government delegates always had a greater capacity to influence the outcomes than any others. It was an artefact of the UN process that the people negotiating the final text on any negotiated decision were government people. Another similarly active interviewee agreed, noting that although UNFF had mechanisms for the participation and engagement of non-country members, it was still a 'country dialogue'. In terms of non-state input, once the discussions went into pure negotiation mode the dialogue between the governments and NGOs and private sector stakeholders, or other international agencies was basically over. There was no further exchange of experience, and very little any stakeholder or any international agency could do from that point on. One NGO informant commented on the impact of the democratic failings of UNFF. Representatives from the World Bank, or UNDP, for example, had exactly the same experiences to NGOs. They usually left after attending the high-level segments deeply disappointed by the extent of discrimination against non-state interests. Consequently, they now tended to send only low-level staff, and there was no real collaboration with the process. Another described the negotiating process as 'really stupid', because non-governmental participants never got to say anything, and had to resort to sneaking around to governmental delegates and whispering in their ears, in order to get them to ask questions on NGOs' behalf. UNFF-5 had been particularly non-democratic and some governments had pushed to exclude NGOs from UNFF-6 entirely. This had led some NGOs to conclude the process should be abandoned altogether.

Democracy functions as a two-tier system in UNFF. Member states have the right to vote, and non-state interests do not. In the case of the latter group, participatory rights are further differentiated between specialised agencies and other intergovernmental agencies, and NGOs. This second-class status affects the extent to which some non-state interests have exercised the rights they have been given, potentially rendering the system less effective. However, in view of the fact that member states do not exercise the right to vote, and that divergent interests spend much of their time simply blocking one another's interests through strategic use of their de facto veto powers, democracy in UNFF is **low**.

Agreement

Formal agreements in UNFF are referred to either as resolutions or decisions. UNFF resolutions are to be understood as part of an increasing trend to create 'soft law' regarding forest issues and which constitute political, rather than legally binding commitments. Decisions, on the other hand, do not even constitute soft law since they deal only with procedural and administrative matters. A resolution has been described as 'a statement of political commitment that has been agreed by a group of states, but which is not legally binding'.[80] Voting by member states may be conducted by a show of hands, roll call or mechanical means, or by secret ballot. Decisions are made by a majority of the members present. Where no member requests a vote on any proposal or motion, the proposal or motion may be adopted without a vote. Equally divided votes are resolved in favour of the negative. At UNFF, as with IPF and IFF, no matter has ever been taken to a vote, and the Forum functions by consensus (understood as unanimous agreement, not following the ISO definition). UNFF's use of consensus led the writers of the *Earth Negotiations Bulletin* in their analysis of NLBI negotiations UNFF-6 to conclude that 'pursuit of consensus on forest issues at the highest level has produced a document limited by the lowest common denominator'.[81]

Apologists from group three of interviewees explained that the UN system was a very complex process, which was a constraining factor. It was also important to understand that there was a subtle difference between consensus and lack of objection. In UNFF, decisions were made on the basis of both methods. It was very difficult to change the rules just for forests, because the same rules existed for every kind of interaction. If there was no consensus, there were no outcomes; this meant that consensus outcomes were not always the best, but it was better to live with the decisions that were made. Other informants from the same group were less positive. There was a majority in UNFF who were prepared to take action but because UNFF ran by consensus, everyone had to wait for the very last of the minority of countries to say yes. Despite the existence of the appropriate procedures, there was no point exercising the powers of majority voting. Voting against powerful countries anywhere in the UN system led to revenge being taken elsewhere. In addition, a minority of countries continually blocked developments and as a consequence there was no real will to change anything at UNFF. One interviewee had reached the stage where they had given up on arguing for the type of instrument (legally binding or otherwise) and would accept whatever they could use to increase the political pressure for action domestically. NGOs – given their minor role as outside observers – were generally negative. One of the interviewees argued that had never been any 'substantive discussion' in either the IPF/IFF or UNFF sessions.

All the processes had been highly politicised, since for many forums the only issue that had mattered was whether they should oppose or support the LBI. Another commented that negotiations were too long and by the end everyone was exhausted. After days of inserting square brackets and alternative wording people got to the stage where they did not care what went in. At UNFF-4, for example, delegates had discussed indigenous peoples' rights and had developed some good ideas, but negotiating the text had proved to be such a failure that everybody simply agreed to drop the whole idea.

The process of reaching agreement in UNFF is deeply flawed. The official procedures, which are clearly defined and recognised by all parties, are never used, and the actual method used (unanimous agreement) is a source of almost universal frustration. Decisions, which might otherwise be approved by a majority, or even a qualified majority of participants, do not get made. This results in difficult and complex negotiations based around compromise. In such a context 'consensus' is used not as the preferred tool, but as the least bad option: no one is offended, but no one is pleased; veto coalitions are the order of the day. Whilst this may reflect the realpolitik of the broader UN system, agreement in UNFF does not merit a rating of anything other than **low.**

Dispute settlement

Dispute settlement mechanisms across the UN system have been identified as inadequate.[82] One problem identified in the literature on multilateral environmental agreements in particular is that dispute resolution mechanisms are often weak, and parties reluctant to use them.[83] Commentators blame the conflict in UNFF on the focus given to text negotiation rather than collaborative dialogue. The emphasis on negotiating text as the primary method of deliberation stems largely from arguments for or against a legally binging convention, and has resulted in a process- rather than output-oriented focus in UNFF.[84] The current method of intergovernmental bargaining has resulted in a form of 'disconnected politics' lacking in coherence.[85] Substantive agreements in comparison increase with greater acknowledgement of diverse actors.[86]

Informants did not comment in detail on the settlement of disputes in UNFF, although the comments made were generally negative. One NGO interviewee characterised most of the UNFF discussions as being one single dispute over national sovereignty, expressed through the conflict over what type of forest instrument to adopt. This was corroborated by a second, who portrayed UNFF-5 as being a 'fight' between the EU, which wanted an LBI, and Brazil and a number of other southern countries, which wanted no legal obligations but plenty of funding. Criticisms were not confined to the oppositional role played by recalcitrant nation-states. One informant from group three also commented on the negative role played by 'professional opponents' in the NGO sector. They functioned at such a high level that they were completely disconnected from the interests on the ground that they were supposed to represent. They were no different from governments, making dogmatic demands that were not really that helpful. Another informant from group three recalled that they did know of one formal complaint. It had been lodged by the UK with the Secretary General concerning the failings of the UNFF Bureau and Secretariat in relation to UNFF-5, and their failure to obtain a formal ministerial declaration at that session. Another interviewee from the same group commented that disputes were sometimes settled through the mediation of countries that were seen as not having a strong vested interest, and were not part of the big power blocs.

Disputes are not settled in UNFF through any formal procedures, and are reinforced by informal, issue-oriented voting blocs, that impede collaborative discussion. With the exception of the entrepreneurial, or go-between, activities of some well-regarded member states, text negotiation appears to be the only mechanism through which disputes are addressed. Dispute settlement in UNFF is **low.**

Implementation

Behaviour change

It has been argued that UNFF – reflecting the practice of other intergovernmental processes – is driven by a set of international environmental norms, which prohibit non-engagement in such prominent environmental issues. No country can be seen to oppose an issue as important as combating deforestation. As a consequence, purely symbolic institutions have arisen, which do not facilitate policy coordination.[87] Indeed, states sometimes even 'deliberately set up "decoy" international institutions to pre-empt governance'.[88] The intention of UNFF is therefore not to provide governance, but to avoid action whilst still appearing busy. Other scholars point out that because UNFF is an entirely voluntary system it 'has made minimal progress on implementation [because] the question of compliance has never arisen, since the IPF/IFF proposals do not actually *oblige* the states to do anything'.[89] The problem with the voluntary approach is that it often fails to deal with intractable, or non-compliant, entities. At the same time, a relationship has been established between 'soft' and 'hard' institutional approaches to changing the behaviour of relevant target groups. The 'harder' the institution the greater is the chance of negotiation improving the quality of implementation and compliance. In softer institutions substantive targets may be watered down during the course of negotiations. In this case, 'intrusive verification and review' are necessary to provide an option for successful implementation.[90]

Attitudes regarding the behavioural change impact of UNFF were mixed. Generally speaking, those interviewees from group three of subjects who actively participated in UNFF were more positive in their analysis. One informant believed that one of the political achievements of the UNFF was that it had succeeded in developing a common understanding of the concept of SFM at a global scale. This common understanding had been further strengthened by the identification of the thematic areas of SFM as part of the deliberations surrounding streamlining the PfA. Another felt that UNFF contributed to changing the political culture and recognition of forests domestically, particularly in terms of the impacts on NFPs. But it was also noted that UNFF did not have any implementation mandate, nor did it play any role in changing fundamental sovereignty issues within countries. UNFF's loss of political status and failure to achieve concrete outcomes had caused political momentum to shift to the CBD and the Climate Convention and it did not carry the same weight as it did when it was established in 2000. The real problem with UNFF was the lack of will by governments to accept the need for a change of policy settings. No amount of increased participation of Major Groups or other interests would make any difference if governments did not change their forest policy perspective. Those interviewees whose role in UNFF was more restricted were also more critical of the Forum's impact, and there was a degree of agreement across the business and NGO sectors that UNFF was not sufficiently changing behaviour at the governmental level. One interviewee for the business sector was dismissive of the 'tired, myopic national forestry regulators' for their failure to get any change at the national level. According to one NGO subject, some countries wanted to write their own rules, and had 'no political will' to change the management of their forests. For many developing countries, the only real motive was to extract money from the developed countries, but there was never any action on the ground.

UNFF's ability to change behaviour relating to forest management is affected by several factors. Analysts identify the subordination of environmental considerations to economic imperatives, resulting in an institution that creates the impression of tackling difficult issues such as deforestation, but does not in fact do anything. This is perhaps an unduly harsh criticism, but it is certainly true that the institution has no implementation mandate, and can do nothing to enforce compliance with any of its substantive outcomes, such as the PfA. Here the fundamentally 'soft' nature of UNFF has a negative impact on behaviour change. This leaves the change impact of UNFF to be largely determined at the national level. Here, the political will – or the lack of it – to change behaviour on the ground is also particularly relevant. Inaction by some member states has contributed to a loss of UNFF's international institutional status, which has in turn further eroded its ability to influence and change behaviour. Behaviour change in UNFF is **low.**

Problem-solving

There are inherent conflicts at play in intergovernmental processes that explain the failure of states to develop regulatory solutions to protect forests. Neo-liberal environmentalism has created a form of global environmental governance, which is beset by ironies and contradictions: it has opened up opportunities for democratic participation, but has also reinforced the role of the market. Such conflicting norms exist uneasily together in practice and result in conflicts that prevent action to protect the environment, most notably the exercise of the precautionary principle.[91] On a more specific level, commentators argue that part of the problem of translating SFM into effective action on a global level, is partly due to a lack of 'political consensus' on who ought to be involved. Within UNFF itself, in contrast to the broader forest policy arena of which it is a part, there has been an erosion of involvement and interest by important forest stakeholders, most notably NGOs and IPOs. This has been attributed to the inability of the IAF to link the international forest policy dialogue to forests on the ground, and consequently the actors most needed to solve forest problems have moved away from UNFF to more innovative and successful initiatives in regions and countries.[92] Consequently, recent years have witnessed an increasing number of multi-stakeholder collaborations outside the UNFF that have developed working relations that have proved more effective in solving problems, such FLEG.[93] These initiatives have been described as 'part of a new system of governance to address illegal logging' that has arisen because UNFF has proved to be 'too slow to deal with the complex issues involved'.[94] UNFF is accused of consisting of a series of 'distant global dialogues' in need of transformation 'into rich, interactive, multistakeholder collaborations that decisively address problems'.[95] UNFF has accepted some of these criticisms. In 2005 Secretariat Director Pekka Patosaari commented on the inability of UNFF to deal effectively with multiple interests and complex problems, and agreed that change was needed to avoid UNFF becoming 'a typical policy talk-shop'.[96]

Views on the problem-solving capacity of UNFF were not sector-specific and were positive, equivocal and negative. Positive views were largely confined to participating delegates, but not exclusively. Looking at UNFF with ten years' hindsight, the positive view argued that the world's forests were better off than without it. A number of advances had been made, particularly in the relationship to the partnerships that had been formed between developing and developed countries. Its role as a reporting mechanism had helped to identify gaps in national forest policies, and recognise what needed to be changed, although it was only one of many catalysts. The UNFF process was more about consensus-building around global priorities and issues. The nature of the global forest problem made a solution inherently difficult. UNFF had to accommodate different interests and values, making discussions more difficult even than climate change. Other comments were more negative. There had now been more than 'ten years of wasteful discussion' and given that the LBI was not going ahead, the negotiations consisted of nothing more than endless discussions with 'no impact on the ground'. Comparisons were drawn with other intergovernmental forest-related processes, including FLEG, UNFCC, ITTO and CBD and non-state processes such as the Forest Dialogue. One interviewee from the business sector argued UNFF should not be defined as a governance system, but 'a debacle', representing nothing more than 'a failed intergovernmental process'. For one NGO interviewee, participation in UNFF had effectively degenerated into 'a kind of damage control' based around forming alliances to block bad positions.

UNFF is one of a number of post-Rio processes that have attempted to grapple with the global forest problematic. There is a strong case that UNFF has provided an enhanced, more universal, intergovernmental forum for dialogue about forest issues, but has not made any substantive new inroads in combating deforestation. Indeed, there are strong arguments to be made that other processes have been more successful in this regard, particularly in terms of addressing the problem of illegal logging. UNFF's problem-solving ability is rated as **low**.

Durability

UNFF has proven to be a resilient entity. At UNFF-6 it was proposed that it would meet every two years commencing after UNFF-7 and its mandate was renewed until 2015. The Forum, however, has had difficulties in developing creative solutions to entrenched policy positions.

At UNFF-5, for example, disagreements over whether the IAF should be legally binding led to public accusations from the EU of inflexibility in the negotiating position of those in favour of the non-legally binding route. This response itself proved unhelpful, and caused negotiations to grind to a halt.[97] On a national level non-state interests expressed a concern during the course of the MSD at UNFF-3 that government structures needed to be more flexible if they were to recognise and take account of the needs and values of forest dependent people at the national forest policy level.[98]

Comments on the durability of UNFF were confined to group three of informants. The fact that it had been going for as long as it had was an achievement in its own right. The Forum did not really generate any ideas of its own, but picked up ideas as they evolved around the world, and simply brought them to the attention of the international community. It did not generate the instruments itself but it certainly gave them a profile; the C&I processes, which were already in existence before UNFF, were one such example. UNFF's originality lay in determining that these processes could be organised around the thematic elements developed within UNFF itself. This approach could be interpreted as either a strength or weakness, of the system. One informant accused the Forum of procedural inflexibility. The MCPFE process was much more flexible. It allowed everybody to speak on equal terms, whether they were government representatives, business and industry representatives, forest owners or environmental NGOs. Everybody had a voice and a stake. In an environment that had so many equally important members, and where each country had an individual vote regardless of forest cover or any other criterion, it was important to have a degree of negotiating flexibility. It was hard for UNFF to reach any form of conclusion when countries were inflexible in their positions.

UNFF has also shown that it can take on the ideas of other bodies and adapt its own substantive outcomes, such as the PfA, to reflect changing policy conditions. The institution has shown itself to be procedurally inflexible, however; it is a victim of intractable negotiations. This militates against an otherwise high performance, resulting in a rating of **medium**.

Conclusions

There is an almost mantra-like quality in the recognition accorded to participation in UNFF documentation. It is frequently placed after the word 'stakeholder' – almost in the same way as 'sustainable' is before

forest management. As with UNFF's predecessors, the term clearly represents an accepted norm of global environmental and forest-related
governance. And yet despite this repetition, participation is a highly
contested concept within UNFF. Interest representation within UNFF is
very mixed. There is certainly extensive state representation and Major
Group interests are involved, but the degree of inclusion is uneven
and on a structural level, non-state interests do not formally participate at all. The MSD is largely tokenistic, since it is extremely limited
in terms of its political influence, especially since UNFF-5. In terms of
resources, only developing countries are provided travel costs, and non-
state interests must generally fund themselves, with some exceptions
being made for environmental NGOs. The institutional arrangements
regarding the accountability and transparency of participation within
UNFF are also problematic, since much of the institution's activities
take place behind closed doors. This may reflect the intergovernmental nature of the Forum, but there are other examples where non-state
interests can listen in on other processes, such as MCPFE.

A second striking feature of the UNFF is that, despite the years of
extensive deliberation, there have been seemingly few substantive outcomes in comparison. UNFF is also marked by few agreements and
frequent disputes. This is attributable to the institutional tradition of
avoiding voting, leading to a kind of 'forced consensus' where the lowest
common denominator predominates in the presence of any opposition.
In this case, the effort to secure an LBI was caught between competing
interests, and in the absence of a vote, fell victim to consensus decision-
making. The apparent lack of any formal dispute settlement processes,
other than those generated by looming deadlines and the fear of failure,
seem to be a feature of negotiations in UNFF.

In terms of implementation, UNFF was created for the express purpose
of putting into practice the PfA first identified in the IPF/IFF processes,
a purpose that it has not been particularly successful in fulfilling and, it
could be argued, has now effectively abandoned in favour of the GOF.
All aspects of reporting, as well as monitoring and assessing individual
countries' performance are voluntary, and there are no means of sanc-
tioning non-performance. This lack of compliance most probably relates
to the absence of enforcement mechanisms, which states have avoided
in favour of voluntarism, and are likely to do so into the foreseeable
future now that a LBI is off the table for the next few years.

UNFF has not been particularly successful in combating deforestation,
which challenges the system's overall impact on changing behaviour. It
could be argued that greater behavioural change has occurred outside

UNFF, in the FLEG processes for example, where several bilateral agreements have been negotiated in an attempt to control illegal logging. Despite its procedural inflexibilities and entrenched power blocs, UNFF has nevertheless proved itself to be a relatively durable institution, and having extended its mandate until 2015, it is likely to remain on the international forest scene for some while yet. UNFF-7 succeeded in keeping the institution alive and there were some slightly more encouraging signs in terms of collaboration at this event than at other sessions. How effective the NLBI will prove itself to be, given its rather weak nature, remains to be seen.

7
Comparative Analysis

Results and discussion

Table 7.1 below sets out the performance of each of the governance systems investigated. FSC is the highest achiever of the four systems investigated, scoring highest in each of the P&C. At the indicator level, it consistently ranks the highest, or joint highest, of the other systems. With a score of 17 out of 25 at the principle level for meaningful participation, FSC achieves 68 per cent, exceeding the threshold value of 50 per cent, and for productive deliberation 22 out of 30, or 73 per cent. With a cumulative score of 39 out of a total of 55, or 71 per cent, FSC's overall performance could be seen as creditable. ISO is the second strongest performer at both the principle and criterion levels. And at the indicator level, it consistently ranks the second highest, or equal second of the systems. With a score of 13 out of 25, or 52 per cent, it just exceeds the threshold required for meaningful participation at the principle level, and scores 17 out of 30 or about 57 per cent for productive deliberation, also placing it second after FSC. With a total score of 30 out of 55, or 55 per cent, ISO's performance could be described as satisfactory. The performance of both PEFC and UNFF is less strong, making them the weakest performers overall. PEFC achieves 26 points overall, or 47 per cent, and UNFF 24, or 44 per cent, meaning that both systems do not meet the threshold value of 50 per cent. Neither PEFC nor UNFF meets the threshold for meaningful participation, with PEFC achieving 10 out of 25, or 40 per cent, and UNFF scoring 11 out of 25, or 44 per cent. For productive deliberation, PEFC meets the threshold value, scoring 16 out of 30, or 53 per cent. UNFF scores 13, failing to meet the threshold, with 43 per cent. In terms of performance at the criterion level FSC exceeds all

Table 7.1 Comparative matrix of governance quality

Principle: 1. Meaningful Participation

Criterion	1. Interest representation Highest possible score: 15 Lowest possible score: 3				2. Organisational responsibility Highest possible score: 10 Lowest possible score: 2			Sub-total (out of 25)
Indicator	Inclusiveness	Equality	Resources	Total	Accountability	Transparency	Total	
FSC	4	3	3	10	3	4	7	17
ISO	3	2	3	8	3	2	5	13
PEFC	2	2	2	6	2	2	4	10
UNFF	3	2	2	7	2	2	4	11

Principle: 2. Productive deliberation

Criterion	3. Decision-making Highest possible score: 15 Lowest possible score: 3				4. Implementation Highest possible score: 15 Lowest possible score: 3			Sub-total (out of 30)
Indicator	Democracy	Agreement	Dispute settlement	Total	Behaviour change	Problem solving	Durability	Total	
FSC	4	5	3	12	3	3	4	10	22
ISO	3	3	2	8	3	2	4	9	17
PEFC	2	3	3	8	3	2	3	8	16
UNFF	2	2	2	6	2	2	3	7	13

	Grand Total (out of 55)
FSC	39
ISO	30
PEFC	26
UNFF	24

four thresholds; ISO meets or exceeds all four criteria; PEFC does not reach the threshold for two criteria (interest representation and organisational responsibility); UNFF does not achieve the threshold for any criteria.

Governance arrangements

Looking at the strongest performer first, FSC most clearly fulfils the requirements for meaningful participation and productive deliberation at a principle level, and the interaction between structure and process appears to be the most collaborative of all four systems in terms of outcomes (behaviour change, problem-solving and durability). In terms of interest representation, FSC has a high degree of inclusiveness, and a satisfactory performance in the other indicators of participation. It is well placed structurally to handle the necessary deliberations associated with multi-stakeholder decision-making. The relationship between structure and process is the most favourably expressed in FSC, since the system also demonstrates a satisfactory level of democracy, and a high degree of agreement. This relationship makes FSC the most collaborative of all four institutions, which may well explain its creditable implementation capacity.

In the case of ISO, the low degree of equality and access to resources amongst participants cuts across its satisfactory level of inclusiveness, impacting on the representation of interests overall, although the institution does exceed the threshold value. A weak performance in terms of transparency also impacts on its level of institutional responsibility as a whole, although the threshold is met, but not exceeded. On a procedural level ISO's weak dispute settlement mechanisms have been the subject of ongoing and unresolved conflict between the more and less powerful interests within the system. If ISO were to be judged solely by its outputs these governance shortcomings might be overlooked in view of the existence of a wide array of globally popular standards. However, in terms of its implementation, ISO's problem-solving capacity is low. Here a link can be made to a lack of performance targets, raising the possibility that accommodating NGO concerns might lead to an overall improvement in problem-solving capacity. In this instance it may be possible to see a correlation between structural and procedural shortcomings and the deficit in the problem-solving aspect of implementation. However, given the collapse of the NGO-CAG Task Force and the hostility of some national standards bodies towards NGOs, prospects for improved non-state participation in ISO are more likely to happen in technical committees other than TC 207. ISO also has a number of commendable points. With 'medium' scores for inclusiveness, resources, accountability, democracy, agreement and behaviour change, and a 'high' score for durability, TC 207 and the ISO 14000 Series generally is in a good position to further improve its performance, requiring only minor structural and procedural adjustments.

Looking at PEFC, the institution is structurally oriented around a restricted set of interests. With such a high level of 'consensus' amongst those involved, it might be argued that higher levels of interest representation and organisational responsibility are not necessary. This argument is reinforced by the fact that the system's decision-making and implementation capacities are passable. On this view, increasing inclusiveness, and other such governance shortcomings, might even prove counter-productive for current participants. But the point should be made that distorted representation has the potential to lead to distorted outcomes.[1] In effect, PEFC, despite the recommendations of the recent review, may continue in its present state in order to maintain itself. But if no change is instituted, opponents to the current arrangements will remain unable to participate meaningfully or make substantive contributions to policy and procedural decisions. Given the restricted set of participants and tightly controlled deliberation, PEFC's implementation capacity will probably remain at its current level of performance, but because it has arisen in such a constrained social–political environment, the system's problem-solving capacity is likely to remain low. If it fulfils the recommendations of its own governance review, and includes a greater range of interests, this situation may improve. In this regard, it should also be noted that PEFC receives 'medium' scores for democracy, dispute settlement, behavioural change and durability. Building on these results, addressing the lower scoring indicators, and following up on the specific recommendations of the governance review will all make an important to contribution to the institution's implementation capacity, which is currently not much below that of either FSC or ISO. Such changes would stand PEFC in good stead for making a substantive contribution to SFM on the ground.

UNFF, although performing at a level similar to that of PEFC, has a slightly different set of structural and procedural issues. It has a more representative and inclusive membership, even if it is only member states that can join and participate as recognised actors. But there are shortcomings in the interactions between participants across a range of indicators, whilst its procedures disenfranchise a number of participating interests. Unlike PEFC, however, UNFF is characterised by a lower level of agreement. In this instance, its more universal inclusion of member states may work against it: unlike PEFC, it cannot simply exclude conflicting interests. Its use of 'lowest common denominator' consensus also results in deliberations that are not particularly productive – especially given the lack of effective dispute settlement mechanisms. It is interesting to note that the three non-state systems perform better in terms of implementation than UNFF. This may be related to the fact that these systems use third-party compliance verification against a set of standards, a model of private rule-making, which, it has been argued, can exceed the regulatory requirements of state-based approaches.[2] An observation might therefore be made that even within voluntary systems, a degree of compulsion is necessary to change behaviour.

If UNFF is unable to improve under its new non-legally binding format, the perception that it is a 'decoy' institution will be reinforced. However, it also needs to be recognised that UNFF functions under a very different set of constraints than the other institutions investigated. Functioning as it does within the UN system, there are some structural and procedural elements that it cannot change. Given these considerations, it is particularly praiseworthy that UNFF received a 'medium' for inclusiveness. Here, the universal inclusion of all UN member states, the MSD, and the recognition accorded to the Major Groups, deserve special commendation. These initiatives have made UNFF far more representative of diverse interests than many other equivalent UN bodies.

Inclusiveness

Inclusiveness is an important indicator as to whether interests are being represented within a given system. Excluding certain interests affects the quality of participation, since there is an insufficient diversity of stakeholders.[3] FSC rated highly under the inclusiveness indicator, and both ISO and UNFF also exceeded the threshold, achieving a rating of medium. PEFC rated low. The three non-state systems use a UN-style 'general assembly' model as their highest organ of interest representation. The FSC's chamber-based structure is the most sophisticated method of including different interests (economic, environmental and social) and although it does not include nation-states on a 'one country one vote' basis it does make specific provisions for including interests on a developed and developing country basis, via the northern and southern sub-chambers. Governmental interests are also included in the economic chamber. Non-members are also well catered for in standards-setting at the international and national levels, if less so at the level of certification assessment. ISO and PEFC do not use the chamber system and include state and non-state interests on a national basis only. ISO appears to have recognised the problem of including interests in this manner, and is beginning to explore new ways of including international NGOs. These two schemes also fall short in the extent to which local interests are included in certification assessment. UNFF, bound as it is by the UN's rules of procedure, is limited in the extent to which it can include non-governmental interests, although it has made some effort to do so. Nevertheless, despite these shortcomings ISO and UNFF, like FSC, are relatively inclusive systems. PEFC is structured in such a way that only a restricted set of interests can participate, excluding others such as international social and environmental NGOs.

Equality

Having a wide diversity of stakeholders is not an end in itself if there is a disparity in status between those within the system. Consequently, inclusiveness needs to be complemented by equality in interest representation. Only FSC meets the threshold for equality. FSC does not score highly, and is perhaps the victim of its own ambitions: it has set itself some lofty targets for gender, social and geopolitical equality in its participatory structures at all levels, which it has not always met. It has also been historically dogged by accusations of undue NGO influence; there is some ongoing resentment amongst business interests, especially those more supportive of PEFC. Mitigating factors are the level of equality enjoyed by non-members in standards-setting and the recognition given to equality by the institution, and its efforts – admittedly not always successful – to achieve it. ISO, PEFC and UNFF are hamstrung by their own governance structures: since each system accords the various categories of interests that participate within them such different status, achieving equality is currently practically impossible. In UNFF national governments share a high degree of structural equality, but this is offset by the inequality amongst other non-state interests. Here inequality is a structural aspect of interest representation, since nation-states have the only procedural authority to participate meaningfully within the system. Even the CPF, despite its delegated powers and collaborative role with other UN instruments, does not enjoy equal status with the nation-states. In both ISO and PEFC structural inequalities are chronic. In the case of ISO, this is partly a historical legacy, since it was established in an earlier era to develop industry-oriented technical standards, which were largely only of concern to business. However, in view of the shift towards more social-environmental standards, the dominant role that economic interests continue to play in standards-setting and executive structures is problematic. The redesign of some of the institution's governance is an acknowledgement of this problem, but is at present anecdotal and too recent to lead to any firm conclusions. To achieve any degree of equality it will be necessary to revisit the whole issue of O-, P- and Liaison (L) members. In the case of PEFC inequality is also structural. At the Board level, economic interests overwhelm social-environmental interests. In the General Assembly there are no arrangements, as with FSC, to counter northern country domination.

Resources

Earlier in this study, a link was made between the availability of resources and the capacity for various interests to participate within a given governance system. There is ample evidence in all the case studies provided that participation in environmental governance is an expensive and time-consuming business. Where there are no resources, or limited resources within a governance system to support participation, those with their own technical or economic capacities clearly have an advantage when it comes to representing their interests. The non-provision, or inadequate provision of resources by a governance system can impact on both inclusiveness and equality. Those who do not have the capacity to represent their interests are effectively excluded. Those with comparatively fewer resources than others have less capacity to wield influence over the system than others. Both FSC and ISO meet the threshold for resources; PEFC and UNFF are rated low. There are some similarities between FSC and ISO in that both institutions have specific arrangements for providing funds for participation at an international level. In the case of FSC, this is a constitutional requirement and covers any member; for ISO, provisions exist under its DEVCO programme, but only for developing country members, not NGOs. Both have rather haphazard arrangements for supporting participation at the national level, depending on the country and the level of resources available to the national body. Neither system provides financial resources for those participating in certification-related consultations. UNFF's funding arrangements are altogether problematic, since a number of its related organs, such as the Secretariat and CPF, do not have funds in their own right, and although there are monies made available for developing countries and some NGOs to attend, these are provided largely by member states on a discretionary basis. In PEFC there is no funding available for international-level participation, and as with FSC and ISO provision of funds for under-resourced interests to participate at the national level is anecdotal.

Accountability

Accountability, often directly equated with responsibility in the literature, is particularly important given the network arrangements and long distance constituencies associated with global governance.[4] Both FSC and ISO meet the threshold for accountability with a medium score, whilst PEFC and UNFF both fail at this indicator level. A common, and alarming theme in the three non-state certification schemes, given the importance of their role, is the accountability problem identified by both commentators and interview subjects regarding the role of certifiers. In all three systems, the relationships between the certifiers, their clients and the bodies whose standards they are implementing have been questioned in view of the potential for conflicts of interest. Part of this disquiet may stem from the structural tension inherent in the role of the certification bodies. Since they are deliberately set apart and independent of the system to which they are accredited, there is at best a weak link in horizontal accountability, raising questions as to whether certification bodies are really answerable to the public. In the case of FSC and ISO this is partly resolved – but also, somewhat ironically, made all the more problematic – by the right for certification bodies to be members of the system. This raises the second problem of vertical, that is internal, accountability. In the case of FSC its own accreditation body has previously sat inside the system, and some ambiguities remain in the current arrangements. ISO certifiers may even play a role in, or chair, committees relating to standards they may subsequently implement. PEFC accredited certification bodies are answerable to an IAF that exists entirely external to the system, but it is almost exclusively accountable to business interests, and certainly not to the public. Accusations have also been raised regarding the undue closeness of some PEFC certifiers to national governing bodies, which they are in turn expected to audit for their eligibility for membership to the scheme.

An argument could be mounted that these certification-related systems are at least horizontally accountable to the wider public through a linkage back to national regulations, and compliance frameworks. Here though, there are problems with the level of stringency in such requirements, and whether the standards under which the various schemes comply are process-, rather than systems-based. ISO and PEFC could be argued to be less accountable than FSC, which either meets, or exceeds national-level requirements, and is performance-based. In terms of vertical accountability, each scheme investigated has a particular set of members who sit closer to the institution itself, and consequently have the possibly to wield more influence than others. In the case of PEFC and ISO these are the national governing/standards bodies, and their business constituents. In the case of FSC, its NGO origins have given civil society groups a greater 'insider' status than others, particularly the WWF. This closeness may have undermined FSC's responsiveness to other 'outsiders' such as European private forest owners.

UNFF does not face the accountability conundrum of how to deal with certification bodies, but as a system, like the other case studies investigated, it has systemic problems. Firstly, its nation-state members are only held horizontally accountable for their actions through periodic, and largely unrelated, national elections. It should also be added than in some national instances such elections are neither free nor fair. Vertically speaking, there are also internal accountability deficits, since member states do not have to answer to other non-member, participating interests, such as NGO participants, or to other aspects of the system (such as the CPF). Here the role of national accountability via NFPs could play a strengthening role, but given the voluntary nature of reporting and implementation, the issue of accountability in UNFF is far from resolved.

Transparency

The more distant processes of governance are from their points of accountability, the more important it is that information is freely available; this is particularly the case where there are no nation-state mechanisms for transparency, such as freedom of information legislation.[5] FSC earned a medium rating, while ISO, PEFC and UNFF failed this indicator with low ratings. In the case of the FSC, the institution's activities were highly transparent in terms of publicly available documentation. ISO is hampered by the restricted release of documentation, and the fact that documentation is released only at certain stages, during standards-setting. The public also faces considerable difficulties simply trying to locate relevant information on a plethora of websites, (such as committee minutes), and the fact that certain information (such as standards, or rules of procedure) must be purchased. In PEFC the 'hollowed out' nature of the institution at a global level means that much information is only available from national bodies, and critical international documents, such as General Assembly minutes, are only available as extracts upon specific request. In UNFF, considerable amounts of discussion between member states occur behind closed doors, and non-state participants as well as the general public are effectively locked out. Publicly available reports are also worded in neutral diplomatic language to ensure the interests of all actors are reflected impartially.[6] This makes it difficult to identify competing points of view, and general observers must therefore rely on external sources, such as the *Earth Negotiations Bulletin*, to interpret events. In these three institutions, transparency in terms of quality and availability of information is problematic. Financial information was not forthcoming from ISO, PEFC or UNFF, which is a problem since running such institutions is expensive, and funds must come from somewhere, such as government, business, or civil society interests, all of whom have the potential to influence institutions behind the scenes if their contributions are not made explicit. In view of the well-established understanding of the relationship between accountability and transparency to 'good' governance in both theory and practice, the scores here leave no room for complacency in any of the systems investigated; this is especially the case for PEFC and UNFF.

Democracy

Both FSC and ISO meet the threshold score of two, or medium rating, whilst PEFC and UNFF both score one point each, or low, thus failing to achieve the required level. Although FSC and ISO have similar results, the reasons are quite different. In the case of FSC, it is fair to say that the institution is highly democratic, and shows a sophisticated blend of consensus and voting. The problem FSC encounters is that it has not been able to fully capture social interests and small forests within its broader decision-making structures. In the case of ISO and TC 207 there are differences in the manner in which democracy is exercised across the different sub-committees, and at times within each sub-committee, effectively franchising or disenfranchising certain participants depending on the democratic mode adopted. In UNFF, although decision-making powers amongst nation-states reside with the least, as much as with the most powerful members, other interests have no franchise at all. In PEFC power is also not distributed evenly amongst voting member bodies; the more production forest in a given country, the more votes. There may be good reasons for this, such as avoiding the decisional gridlock that characterises UNFF, but in both PEFC and UNFF there is a clear democratic deficit.

Agreement

FSC scores the highest rating of three, while ISO and PEFC meet the threshold of medium. UNFF only scores one point, and fails to meet the threshold required to pass this indicator. FSC scores the highest rating possible because of the fact that the system as a whole is characterised by a very high level of agreement between participants over the decisions made. The key to this success is not that the FSC uses 'consensus' on all occasions, but that the stages leading to the making of a decision make agreement increasingly likely. In the General Assembly, for example, a motion does not make it to the floor unless there is a majority in favour of the motion in all chambers, or until it has been amended in such a way as to be acceptable to all chambers, whereupon it is put to the vote. At this point the qualified majority (66.6 per cent) voting system often results in agreement by overwhelming majority. ISO has some of the same outward appearance of using consensus as FSC, but without the same clear procedures as FSC the reaching of agreements can be a confused process. Decision-making switches between different modes of agreement (consensus or qualified majority, or majority) in a relatively arbitrary fashion, and consensus suffers from a lack of clarity over what 'sustained opposition' means in practice. PEFC rates similarly to ISO, but is clearer as to how it reaches agreements on the international level where the standard practice is essentially majority decision-making. Its problems are at the national level: there are inconsistencies between countries over the methods for reaching agreement, and like ISO, it also suffers from a lack of clarity over what is meant by consensus within standards-setting processes. UNFF on the other hand, whilst universally employing consensus (despite being procedurally bound by majority rule), has a generally low level of reaching agreement. Here enfranchised participants are pitted against each other according to which veto coalition they belong to.

Dispute settlement

In all four case studies, the observation that any consensus-seeking process requires that participants enjoy the right to be heard, and have their concerns taken into account, is particularly valid.[7] It is interesting to note that both FSC and PEFC have made some effort in recent years to address the shortcomings in their dispute settlement mechanisms, although the success of these changes is as yet unclear. In ISO there appears to be an institutional preference for settling disputes informally. The right to engage in dispute settlement procedures is limited to participant members, and unsettled matters simply pass up the chain of command until they reach the Council, which has the final say. ISO is therefore hampered by a lack of formal avenues for appeal in some specific aspects of the system (such as membership accreditation and conformity assessment). In both cases neither formal nor informal measures have assisted in settling protracted disputes. In PEFC it should be noted that nationally, the emphasis is more on the existence of dispute settlement mechanisms than whether they actually settle disputes, since concerns expressed by interested parties only require consideration. In UNFF disputes are not subjected to any formal procedural mechanism and tend to be manifested instead in the inflexible positions adopted by the various nation-state factions within the system.

Behaviour change

It is interesting to note that all three of the private governance systems meet the threshold for behavioural change, whereas UNFF does not. Business subscribers to each of the three programmes are required to modify their behaviour, if they have not already done so, to meet the systems' requirements, and maintain it, if they wish to retain their certification status. This is not to say that there are not substantive differences between the systems investigated – not least of which is the discrepancy between performance- and process standards – but all three internalise certain expected norms of behaviour, such as continuous improvement, for example (even if there is variability as to whether improvement is monitored). The extent to which participants in each of these systems have also changed behaviour as a result of mutual learning is present in each programme, although the extent and type varies. In FSC, for example, there has been a high degree of inter-organisational learning resulting from its multi-stakeholder structures and this is present to a more limited extent in ISO also. In both instances, although to different degrees, interaction and communication have engendered social learning. PEFC participants have opted for a narrow, technical response to changes in the forest policy environment; the system has a relatively low capacity for social learning on account of its exclusive structures, and having filtered out dissenting viewpoints, has restricted its scope for policy learning.[8] With its strong scientific orientation and intergovernmental policy focus UNFF's capacity for social learning is also restricted; governments act as gatekeepers of the policy agenda and largely serve their own political and economic agendas, whilst other interests play a largely subordinate role.[9] Non-state agents may be the biggest potential agents for change, but UNFF's procedures limit their role, and the institution's ability to change behaviour is limited. It has neither compliance nor sanction mechanisms and relies on a system of voluntary implementation and reporting. Here the argument that a lack of strong norms and seriousness concerning implementation may be decisive factors in a regime's effectiveness may be particularly relevant for UNFF.[10]

Problem-solving

FSC achieved the highest rating of medium, meeting the threshold for this indicator, but the other three systems only achieved low ratings, thus failing to meet the threshold. Such weak performance in all of the systems should be a cause for concern. It is perhaps worthwhile remembering at this point that all four systems arose in the context of the UNCED, where sustainable development was identified as a major objective of *Agenda 21*. In terms of forest issues, *Agenda 21* was clearly aimed at combating deforestation, but the means of achieving this objective came to be increasingly couched, particularly through the IPF/IFF processes, in terms of SFM. It has been pointed out, however, that this concept varies according to which groups of actors and interests define it.[11] FSC chose to avoid the idea of SFM as a normative concept and opted instead 'to promote the responsible management of the world's forests'.[12] PEFC specifically refers to its role as being 'to promote SFM'.[13] UNFF's aim is 'to promote the management, conservation and sustainable development of all types of forest and strengthen long-term political commitment to this end'.[14] The initial impetus of the ISO 14000 Series, arising out of Rio, was to address clean production and tackle hazardous waste. This objective was ultimately transformed into a vision and a philosophy that environmental management standards 'provide an effective means to improve the environmental performance of organisations and their products, facilitate world trade and ultimately contribute to sustainable development'.[15]

FSC, PEFC and ISO have opted to address the 'problem' of deforestation through the 'solution' of sustainable development in different ways, and with divergent strategic objectives. In the case of FSC, the original aim of addressing deforestation by tackling the unsustainable harvesting of tropical timber has been partially obscured by a greater level of uptake in the arguably already well-managed forests of Europe, and the developed world generally. For PEFC, two underlying motives were the desire to create a market alternative to FSC in Europe (and subsequently the world), and provide a locus for forest owners outside the FSC system. It could be argued that SFM was a secondary, more public, market objective. Being non-state initiatives, both PEFC and FSC also have their own sets of external constraints, the most notable being their market-based orientation, and their inability to impose compliance on forest actors in the same way as nation-states might do. It is also worth considering whether market-based instruments are the right tools to tackle deforestation, since it has been argued that deforestation is a consequence of current market ideology. Public goods (forests) have been converted to private assets in a global neo-liberal economy that promotes voluntary regulation, a policy approach that inherently renders action ineffective.[16] Furthermore, an argument has been advanced that certification is simply too narrowly focused for tackling such a comprehensive problem as deforestation.[17] But given that non-state regulatory mechanisms now seem an accepted complement to tackling problems when the state proves insufficient, there is a strong likelihood that non-state measures will continue into the future.[18]

Despite being composed of nation-states with potentially more regulatory power than such non-state initiatives, UNFF's goal to promote sustainable management has also been restricted, in its case by political and diplomatic considerations. This has resulted in a voluntary and non-legally binding system, which has been effectively stripped of any enforcement capacity. Whether this will assist or hinder action on combating deforestation is not yet clear. It should also be remembered that all non-state systems passed the implementation threshold, while UNFF did not. This may be a peculiarity of UNFF, since it has no effective implementation capacity by design. Given its massive mandate, and broad-based membership, if UNFF had made strong agreements backed by powerful sanctions it might have had a more significant impact on the ground. Other intergovernmental processes are legally binding, such as the EU's Forest Law Enforcement, Governance and Trade Action Plan. This has arisen in the wake of UNFF as an alternative mechanism with a focus on illegal logging, and demonstrates the value of an ongoing role for governments in combating deforestation.

The fundamental question arises from these case studies as to whether a system that fails to tackle a given problem can be considered to have succeeded either in terms of implementation, or more broadly, as a governance system. A preliminary conclusion offered here is that this depends on the extent to which a system has passed or failed in other areas of governance; that is whether it has the other necessary structural and procedural components in place to improve its performance. These components are important since, given the complexity of the issues surrounding deforestation, it should be recognised that combating such a problem is likely to take time, and the longer-term efficacy of an institution may depend on the quality of its governance arrangements overall.

Durability

All four systems passed the threshold, FSC and ISO both rating highly, and UNFF and PEFC achieving a medium rating. The FSC has built in the ability for its regional standards to be sufficiently flexible to fit into local conditions, whilst at the same time meeting international accreditation requirements. It has also demonstrated its ability to adapt to changing market conditions (most notably the arrival of competitor programmes) and more recently in ensuring that the contents of its standards are mutually consistent. Some criticisms have been levelled against the institution: it has been slow to adapt to the needs of its business constituents and its standards vary in their stringency both on a country-by-country level, and in terms of generic versus national standards. ISO reflects a similar ability to adapt to changing market conditions on an even broader level, moving in a more historical context from product- to process-based standardisation, including an increasing range of social–environmental standards beyond EMS, such as social responsibility. But there has been some institutional resistance, and the development of these new standards – in TC 207 at least – has also demonstrated a degree of inflexibility in responding to the arrival of new interests and their needs. TC 207 has also been criticised over consistency, as there are large discrepancies in environmental performance between different companies, depending on the stringency of their (self-determined) environmental objectives. UNFF shows a degree of institutional adaptability, by absorbing much of the intergovernmental forest arrangements from IPF/IFF as well as the various C&I processes. It has also demonstrated a degree of flexibility in reinterpreting its mandate: the decision, after much fruitless debate, to abandon a LBI, and instituting the GOF are such examples. PEFC has demonstrated an ability to adapt to changing market conditions, particularly in moving from a European to global programme, but this may have been a matter of pragmatic expediency. The institution has a tendency towards excessive flexibility in terms of its national standards, which can be either performance- or systems-based, and either nationally or regionally implemented. This raises questions over the consistency of its standards.

Institutional typology

The Introduction questioned the adequacy of traditional institutional governance methods to the task of managing forests in the current era. Such a claim required some justification, and an examination of the theory and practice of governance noted firstly a shift away from top-down, to more participatory approaches, and a transition to deliberative, as opposed to aggregative, democratic modes. Three factors, or parameters, were identified as influencing the institutional expression of contemporary global governance: authority (state versus non-state), democracy (aggregative versus deliberative) and innovation (new versus old). In their respective case study chapters, each institution was classified on the basis of these parameters, and the quality of its governance evaluated.

FSC's structure and process would appear to combine to make the institution more suited to the social and political needs of current era of globalisation than the other institutions selected. Its institutional type also contributes to its success. Figure 7.1 below locates each of the case studies in the model of institutional classification presented in Chapter One. The first point to note is that both UNFF and PEFC, the weakest performers of the four case studies, occupy the portion of space that is more oriented to the state-centric end of the authority (x) axis, a situation repeated on the aggregative end of the democracy (z) axis. It is interesting that they fall on opposite sides of the zero-point for the innovation (y) axis. The similarity of their overall performance makes it difficult to be conclusive, but it is interesting to speculate as to whether

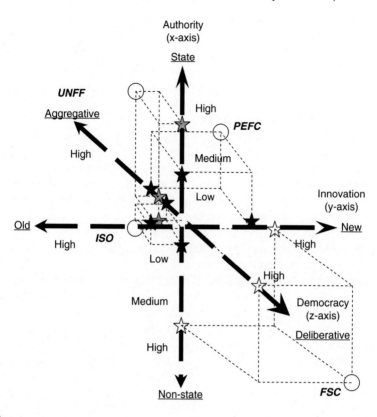

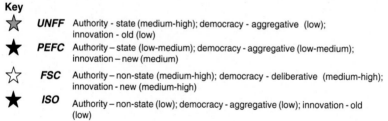

Key

☆ (grey)	**UNFF**	Authority - state (medium-high); democracy - aggregative (low); innovation - old (low)
★	**PEFC**	Authority – state (low-medium); democracy - aggregative (low-medium); innovation – new (medium)
☆	**FSC**	Authority – non-state (medium-high); democracy - deliberative (medium-high); innovation - new (medium-high)
★	**ISO**	Authority – non-state (low); democracy - aggregative (low); innovation - old (low)

Figure 7.1 Institutional classification of governance type: UNFF, PEFC, FSC, ISO.

PEFC's more innovative governance style has contributed to its marginal lead over UNFF.

More significant, however, is the possibility that poor overall performance of these two institutions is linked to their state-centric, aggregative–democratic orientation, which raises some potentially profound questions regarding authority and democratic practice. Does this mean that institutions with a greater nation-state orientation and traditional approaches to democracy are not suited to the social and political imperatives of contemporary global governance?

The second point of interest is that both FSC and ISO, the higher performers, provide a mirror opposite to UNFF and PEFC, occupying the non-state end of the authority continuum. In terms of innovation, ISO sits slightly along the 'old' end of the y-axis, while FSC is much further along the 'new' end of the axis. A similar situation exists on the democracy axis. Bearing in mind the discrepancy between the two in terms of overall performance, it is worth noting that there is a greater level of divergence between both institutions regarding their location on the axes for innovation and democracy. Here it may be possible to infer some relationship between democracy, innovation and overall performance. There may also be a link to their generally higher performance and the non-state nature of their authority.

Looking at FSC, the most highly rated governance system of the four case studies, it is clear that it sits closest to the non-state, deliberative, and 'new' ends of the authority, democracy and innovation axes of any of the case studies. It should also be noted that the FSC is the only institution clearly placed on the deliberative end of the democracy axis. This may lend some credence to the linkage made in the Introduction between the 'fit' between deliberative modes of democracy and contemporary global governance, in contrast to the inherent conflict between economic, environmental, and social interests participating in governance systems that adopt a more competitive, aggregative–democratic approach.

8
Conclusions

Case lessons

This study has looked in some detail at the construction of global governance, making use of the environmental domain as a thematic area in which to investigate its contemporary expression in finer detail. Generally, it has argued the case that participation and deliberation are integral to the structures and processes, which undergird legitimate governance in contemporary global institutions. Specifically, using a detailed framework of governance-related PC&I for evaluating governance quality, it has investigated a range of systems, and has assessed their performance. This framework has gone beyond the relatively random selection and application of criteria used to determine governance quality elsewhere. Having applied the framework to the forest policy arena, clear and specific differences between governance systems have been revealed. The ways in which these systems differ means that the institutions that have been investigated are not to be understood as similar entities: once an analysis of governance is applied consistently, as it has been done here, it is possible to see how these differences impact on the legitimacy of the institutions in question. If the institutions investigated are interested in 'best practice governance', and wish to address their legitimacy deficits, they should take a closer look at the indicators in which they fall short. It is hoped that the study that has been undertaken here will encourage institutional participants and general readers to ask their own questions as to how democracy is being, and can be, practised to meet the needs of the third millennium.

Each of the case studies has also been further delineated as representing a specific type, and the case has been made that there is a relationship between this classification and overall performance. Contemporary governance is expressed in a number of divergent, and at times, competing, models. Whilst all may share similar policy objectives, and at times provide complementary approaches, they may vary markedly in their outcomes. This may apply even within apparently similar regulatory models such as certification, and are reinforced, in this case, by the rivalry between market competitors. Such rivalries point to a tension within the market-driven emphasis in contemporary governance, and especially within the discourse of sustainable development. This discourse has been adopted, and indeed, it has been argued, aggressively pursued, through a particularly dominant ideology of the contemporary era, neo-liberalism.[1] But as the Introduction noted, the cornucopian model of limitless economic growth in a finite world has itself contributed to environmental degradation. The tensions being played out in forest certification may in fact represent a more profound contest over the definition of sustainable development itself.

And so it is that the global environmental policy arena is built upon the conflicting values of economic rationalism on the one hand and the need for environmental protection on the other. How to collaborate in such potentially fraught contexts is one of the most important challenges to this rapidly evolving domain of global policy-making. The increasing role played by the private sector, NGOs and other non-state actors at all levels has necessitated the development of alternatives to the traditional methods utilised by nation-states. Environmental governance therefore typifies a contemporary trend for interaction between decentralised networks made up of multiple actors.[2] The institutions in which these interactions occur are manifold, and even within a single policy domain there are multiple approaches to addressing common problems. Although 'governance without government' remains someway off, there is nevertheless a wide array of mechanisms of social–political decision-making, ranging from the centralised and hierarchical, to the decentralised and self-regulatory.[3]

As has been already, forest governance provides an ideal laboratory to scrutinise the nature of economic and political trade offs in different types of global institution. Where a particular policy arena is built around conflicting values, the role played by governance to address the particular issue becomes a critical one. So too does the style of collaboration utilised to negotiate, make and implement decisions addressing the policy issue. Where there are value conflicts inherent in the policy area, a deliberative approach encompassing multiple sets of interest, as demonstrated in this book, performs better than one that seeks to serve only single or dual interests, since this only entrenches conflict.

In the case studies investigated, it is possible to draw two significant conclusions regarding deliberation and conflict. Firstly, the four institutions place different emphases on certain sets of interests. In the case of UNFF, governments are to be identified as the single interest grouping, with the greatest access to decision-making. PEFC clearly favours forest owners and the forest industry. These two institutions impede the participation of a range of interested parties in substantive discussions, and can be interpreted as being either single- or dual interest focused. ISO has a largely business/technocratic emphasis, although this is now being challenged by non-state interests. FSC has a legacy of close relations to environmental NGOs and has had some problems with the representation of social interests, but as a system, it has a broad-based constituency. It is the one system in which genuinely 'multi-stakeholder' deliberation is central to decision-making processes.

Secondly, given the variation in the performance of the institutions investigated, it appears that the practice of democracy is not yet at an optimal level to adequately account for the multiple actors involved in global forest governance. The lack of consistency across the case studies should be a source of concern as it indicates that in terms of their broader structures and processes, some of the institutions investigated are less democratic than others. This has implications for their legitimacy, and therefore, the value to some interests of participating within them at all.

How the implications of this study will be taken up by the institutions investigated are for them to decide; it is possible that they may wish to make improvements in their governance as a result. For them, and other institutions that may be interested in evaluating their own performance, determining if a minimalist or elaborate programme of improvement is necessary to achieve gains in performance would be a practical area for further research. A minimalist approach, for example, might look at making improvements at the individual indicator level, whilst a more elaborate programme would entail changes at the criterion and principle – i.e. at the broader structural and procedural – levels. For the FSC the changes required would appear to be at the indicator level. In the case of ISO, the changes required would appear to be slightly more significant, but also relate largely to the indicator level. For PEFC, which this study has presented as being relatively state-dependent despite its non-state orientation, it might be a question of increasing its autonomy from the state, shifting its governance away from a 'state democratic' model, and making other changes to its structures and procedures.[4] In the case of the intergovernmental programme UNFF the degree of change necessary will mean according non-state interests a much greater role, and moving beyond the rules and procedures of the UN system as they stand. Such changes may require a degree of change at the principle and criteria levels that these latter two institutions neither can, nor will, implement. In the case of UNFF the task would seem insurmountable, as the whole UN system would need to be similarly reformed. However, given the analysis in this book, some changes are essential – at the indicator level at the very least.

It is difficult to draw any definitive conclusions from only four institutions, however. This book should therefore be seen as developing some insights into, and contributing to a revision of, some of the theories of contemporary governance, but not definitive in its own right. If the findings from these studies are to be used to argue that there is a causal relationship between participation, deliberation and quality of governance, a greater number of case studies across a wider range of other institutional types will be necessary to determine if the trends identified here are correct. Further research would also be useful to determine whether market-based governance systems are more, or less, effective than other models, and which specific market models have the greater problem-solving capacity. Limitations notwithstanding, the insights gained from the four case studies have a wider relevance beyond forests. Social responsibility, for example, including ISO's own 26000 Series, would be a suitable candidate. Fairtrade and other commodity labelling programmes, such as the Marine Stewardship Council, or the Extractive Industries Transparency Initiative, also spring to mind. Research into the framework's applicability to climate change management and associated and market-based systems for emissions trading might yield interesting results. Given the focus on responsible investment in such markets, an investigation into the governance quality of these new and emerging financial mechanisms, which underpin them, could also be useful. All these examples would certainly provide deeper insight into the practice of sustainable development in the current era.

The methodology adopted in this study implies that consistently formulated hierarchies of PC&I have the potential to be applied at all spatial levels.[5] It also allows for the creation of standards that can serve as a reference for monitoring, assessment and reporting across these levels.[6] It would be entirely possible to develop quality of governance standards out of the framework used in this study that could be applied at the global, national and local levels. In view of the inconsistency of the governance arrangements utilised by each of the case studies investigated, the disagreements in the literature over the various attributes of governance, and the 'self-certification' currently in place, such meta-standards are in fact essential. As the world comes to grips with a range of global problems, and social–political interactions increasingly shift to non-state democratic contexts, governance standards will become the main means by which governance legitimacy can be assured. Quality of governance standards will make it easier for potential participants to determine whether they should engage in a given process or not. It will avoid the uncertainty that currently exists over the legitimacy of a given system, and whether to lend it credibility by participating.

Review of research methods

There are some aspects of the methodology developed in this study that require further refinement. Although the analytical framework adopted has been consistent across institutions the means by which their performance has been assessed is less so. The governance issues confronting any system are unique but it might still be useful to extend performance evaluation to a finer level of detail to include generic verifiers (or sub-indicators) for each indicator.[7] This could provide a uniform set of data inputs against which each indicator could be evaluated.[8] In the case of transparency for example, one verifier might be the public availability of financial data (not readily available in three of the four institutions studied here). Secondly, at present each indicator is equally weighted within the relevant criterion. In the case of interest representation, for example, this places the same degree of significance on inclusiveness, for example, as it does on the provision of resources. The value of this approach needs further investigation, as it has been noted that the scoring and weighting of indicators and determining their relative importance is a subjective exercise.[9] Certain verifiers would also be sector specific, however. Measuring behaviour change in the forest sector would be different from that expected in climate change management, even if there might be some overlap (such as reducing emissions) and these would be most logically developed through standards-setting processes.

With hindsight, there are two principle shortcomings with the analytical methods used in this book. The key informant interviews were

qualitatively analysed, meaning that the author exercised a considerable degree of discretion. Removing the investigator from the process,
and developing a method of quantitative assessment would allow for
a level of inter-subjective evaluation not present in the existing case
studies.[10] Secondly, collecting the data associated with the case studies
took several years. The results presented in each of the case studies
imply that evaluation was static, occurring at one point in time.
This was not in fact the case. During the course of research, some
governance arrangements were changed; with FSC and PEFC new dispute settlement mechanisms were put in place. In the case of ISO,
TC 207's NGO-CAG Task Force – which looked like it would deliver
some significant governance changes regarding interest representation –
collapsed. For UNFF, the agreement to pursue the NLBI and the creation of the GOF occurred after the main data acquisition phase. All
these developments were subsequent to the key informant interviews.
A way to address this problem in future would be to make evaluation
'static'; this would indeed allow for different 'cuts' into the institution over time, to track changes in governance arrangements.[11] In
addition to tracking changing perceptions regarding governance quality over time, it would also be possible to compare multi-stakeholder
attitudes, for example, between governmental (or state) and non-
governmental (or non-state) interests, more widely across sectors, or by
region (i.e. North/South).

With these observations in mind, a re-designed model for evaluation, which gives greater voice to the participants in the governance
system under investigation, is presented below, focusing on the UN
initiative aimed at reducing emissions from deforestation and forest
degradation in developing countries (REDD). This initiative combines
nicely the issues of multi-level, multi-stakeholder governance discussed
in this volume. REDD addresses the problem of climate change via a
range of state and non-state market-based mechanisms to encourage
sustainable management of tropical forests (or SMF), and thereby reduce
greenhouse gas emissions. It is now formally referred to as REDD-plus
in the wake of the United Nations Framework Convention on Climate
Change (UNFCCC) Conference of Parties (COP) 15 in Copenhagen, to
reflect the initiative's growing emphasis on conserving and enhancing forests on the basis of their value for carbon sequestration, rather
than simply reducing emissions.[12] The '-plus' in REDD-plus widens
the scope of the mechanism to include conservation and enhance-
ment of forest carbon stocks, as well as SMF. This means that activities
such as improved management of protected areas, forest plantations

and restoration, and reduced impact logging may yet be elements of REDD-plus strategies. The definition of SMF, an interesting terminological development, and specifically how it will be distinguished from 'SFM' is not yet clear. However, the definition is certain to cover many of the community-based forest management practices undertaken by local communities and Indigenous peoples. The broadening of REDD to REDD-plus is generally seen as a positive move.[13]

As there is no final and binding REDD-plus agreement, nothing in the COP-15 draft text can be described as certain. However, negotiators at COP-15 did reach consensus on a number of key issues, which are extremely likely to be part of a REDD agreement when it is reached. There is still ample opportunity for forest sector stakeholders to influence REDD-plus negotiations to ensure progressive and equitable outcomes benefiting both people and forests.[14] There are three principle REDD-plus related mechanisms: the UNFCCC, responsible for the intergovernmental negotiations regarding the content and format of REDD-plus; The United Nations Collaborative Programme on Reducing Emissions from Deforestation and Forest Degradation in Developing Countries (UN-REDD), which is supported by UNDP, FAO and the UNEP and manages the technical and financial components of the initiative at the international and national level; and The Forest Carbon Partnership Facility (FCPF), which via the World Bank, provides funding aimed at maintaining standing forests by encouraging biodiversity conservation and sustainable use through a range of country-level projects. National governments and NGOs such as The Nature Conservancy provide funds for the initiative.[15]

Below are the results of two anonymous surveys of governmental and non-governmental attitudes amongst participants in UNFCCC REDD-plus related negotiations, conducted in November 2009 (before COP-15; in brackets) and March 2010 (after COP-15). Participants were asked to rate their perceptions anonymously, by means of the Internet tool SurveyMonkey, using a Likert scale from 'very low' to 'very high' (1–5), rounded to the second decimal point. The indicator ratings were added to produce a result at the criterion level, and the relevant criteria added to determine principle scores; finally, the two principle scores were combined to determine overall performance (see Table 8.1). In order to compare perceptions respondents were stratified into four subgroups: (1) environment North; (2) environment South; (3) government North; and (4) government South. Using standard statistical methods, the average ratings of each of the four sub-groups were in turn used

Table 8.1 UNFCCC-REDD-plus related questionnaire – government and environmental NGOs – North and South

Principle	1. Meaningful participation Maximum score: 25; Minimum: 5				2. Organisational responsibility Maximum score: 10 Minimum: 2			Principle Score
Criterion	1. Interest representation Maximum score: 15 Minimum: 3							
Indicator	Inclusiveness	Equality	Resources	Criterion score	Accountability	Transparency	Criterion Score	
Environment	2.67	1.83	1.00	5.50	2.60	2.67	5.27	10.77
North 6 (19)	2.58	1.83	1.59	6.00	2.76	2.76	5.52	11.52
Environment:	3.25	2.67	2.17	8.09	3.46	3.36	6.82	14.91
South 17 (5)	2.60	3.00	1.80	7.40	2.00	2.40	4.40	11.80
Weighted	3.10	2.45	1.86	7.41	3.24	3.18	6.42	13.83
average	2.58	2.07	1.63	6.29	2.60	2.69	5.29	11.58
Government	3.33	4.00	1.00	8.33	3.00	3.67	6.67	15.00
North 3 (1)	5.00	4.00	1.00	10.00	3.00	4.00	7.00	17.00
Government	3.50	3.10	2.33	8.93	3.30	2.80	6.10	15.03
South 10 (5)	3.20	2.60	2.20	8.00	3.25	3.20	6.45	14.45
Weighted	3.46	3.31	2.02	8.79	3.23	3.00	6.23	15.02
average	3.50	2.83	2.00	8.33	3.21	3.33	6.54	14.88
Combined weighted	3.23	2.76	1.92	7.91	3.23	3.12	6.35	14.26
averages	2.77	2.23	1.71	6.70	2.72	2.81	5.54	12.24

	3. Decision-making Maximum score: 15 Minimum: 3			2. Productive deliberation Maximum score: 30 Minimum: 6	4. Implementation Maximum score: 15 Minimum: 3				Principle Score	Total (out of 55)
Democracy	Agreement	Dispute settlement	Criterion Score	Behavioural change	Problem solving	Durability	Criterion Score			
2.20	2.17	2.20	6.57	2.60	2.60	2.67	7.87	14.44	25.21
2.33	2.29	2.15	6.77	2.83	2.53	3.44	8.80	15.57	27.09
2.71	2.88	2.54	8.13	3.06	2.94	3.38	9.38	17.51	32.42
2.25	2.00	2.00	6.25	2.60	3.00	3.00	8.60	14.85	26.65
2.58	2.69	2.45	7.72	2.94	2.85	3.19	8.99	16.71	30.54
2.31	2.23	2.12	6.66	2.78	2.63	3.35	8.76	15.42	27.00
3.33	2.67	2.33	8.33	3.67	3.67	3.00	10.34	18.67	33.67
4.00	4.00	4.00	12.00	4.00	4.00	5.00	13.00	25.00	42.00
2.80	3.40	2.78	8.98	3.70	3.50	3.60	10.80	19.78	34.81
3.20	3.00	2.50	8.70	2.60	2.40	2.80	7.80	16.50	30.95
2.92	3.23	2.68	8.83	3.69	3.54	3.46	10.69	19.52	34.55
3.33	3.17	2.75	9.25	2.83	2.67	3.17	8.67	17.92	32.79
2.70	2.89	2.53	8.12	3.21	3.10	3.29	9.60	17.73	31.99
2.52	2.42	2.25	7.18	2.79	2.64	3.31	8.74	15.92	28.16

to calculate the weighted averages for the two main groups (environment and government). In order to ascertain the overall perceptions of all respondents, combined weighted averages were subsequently also evaluated and compared, thus articulating a cross-sectoral 'consensus legitimacy rating'. Participants were recruited variously, either from publicly available Internet lists of named individuals active in UNFCCC or otherwise named as being associated with REDD-plus to some degree. An initial cohort of approximately 200 valid email addresses was generated by these means. A total of 36 individuals completed the second survey (30 for the first). Individual recipients were encouraged to forward the invitation to participate to their colleagues; it is not known how many participants responded as a result of secondary recruitment into the survey. The presence of complete data sets for environmental NGOs and government (global North and South) provides a useful combination of perspectives, which, it could be argued, represent two 'poles' along the governance continuum (non-state/state).[16]

The point to be stressed is that the intention here is not to provide a detailed commentary on the governance of the UNFCCC REDD-plus related negotiations, but rather to illustrate the potential that the revised analytical framework provides for both determining and tracking multi-stakeholder attitudes regarding institutional governance. It should also be added that this survey is not to be seen as being especially representative of the whole policy community engaging in REDD-plus related negotiations. But a few general comments can be made. The most interesting result is the overall improvement in the rating given by combined ENGOs. This would appear to reflect historical events. Whilst other negotiations were not overwhelmingly successful, the REDD-plus related discussions at COP-15 were relatively productive, with NGOs enjoying a fairly high degree of access to text negotiations, even if they remained formally outside the deliberations themselves.[17] This perception is apparent in the inclusiveness indicator, although it is also possible to note an increase at the criterion level for organisational responsibility, and transparency in particular. Government also shows increased confidence in the mechanism's implementation capacity, in contrast to the overall failure of the Conference, and the consensus legitimacy rating of both sectors (state/non-state) is higher. It is also worth noting that both groups provide a consistent low score for the resources indicator, which is clearly the 'poor relation' of UNFCCC's governance arrangements. Finally, one further interesting feature of both surveys is the generally higher rating given by the global South – both governmental *and* environmental NGOs. This might seem to indicate that as an initiative '*for*' the South some of the traditional North/South imbalances are reversed.

Advances in governance theory and practice

Much of the research underpinning this book was undertaken between 2004 and 2008, and there have inevitably been some developments in governance theory subsequently. Of particular interest to this study is the ongoing preoccupation with the authority and legitimacy of global institutions seeking to address conflicts around global public goods, and the role of state and non-state actors in such contested spaces.[18] The representation of non-state interests continues to be stressed as an essential component in addressing the participatory deficits at the global level, and an essential element of input legitimacy. There is also recognition that more inclusive global institutions have a role to play in addressing some of the governance gaps that currently exist.[19] It is interesting to note that such institutions are identified as contributing to a more *heterarchical* – as opposed to hierarchical or anarchical – conception of global governance, which more 'adequately describes the dense web of international governance institutions created and maintained by public and private sector actors'.[20] This observation is highly pertinent to forest governance, and would extend the discussions in Chapter Two that the many different institutions addressing forest management at the global level collectively constitute, perhaps, less of an international forest 'regime', than a heterarchy.

Forests also continue to be viewed as a contested public good in the literature. It is not the 'privateness' or 'publicness' of forests that determines this (especially as ownership is constantly vacillating between public and private hands as neo-liberal or social-democratic land use policies predominate). It more a case of societal choice, and this is affected by the 'globalness' of a particular good.[21] There is an argument emerging that governance, especially participatory governance, can assist in the creation of mutually beneficial outcomes, as it can reduce controversy and facilitate 'durable policy consensus'.[22]

One prediction is that where private goods and private actors intersect with public goods, market-based mechanisms and standards have a higher chance of reducing controversy than those generated by intergovernmental agreements.[23] An important caveat is to be attached however:

> Such private standard setting can hardly be considered in instances where the interest of individuals and groups, who cannot fully

participate in such rule-making exercises appropriately, are concerned. In view of the new enthusiasm for private rulemaking within some parts of academia, it should be noted that proper stakeholder involvement is an essential precondition for private activity in this regard.[24]

This observation lends weight to the perspective in recent governance theory that institutional trends towards 'partnership and dialogue', have a potential to overlook the reality of contestation as a central aspect of governance. Such evolving modes still need to address asymmetries in power relations by respecting democracy and accountability and promoting genuine participation. Here 'multi-scalar' and multi-stakeholder initiatives (MSIs) have something to offer.[25] But even in this context issues of structural inequality remain due to the inadequate resources available to some actors, and caution should be exercised before attributing democratic legitimacy to such projects.[26] This is seen as being explicitly the case in governance of global commodity chains.[27] Timber is of course one such commodity.

The international political economy is clearly witnessing a proliferation of business models with a greater level of social responsibility than before, although these should still not be interpreted as representing the prevailing global norm.[28] There also appears to be an emerging competition between older models of corporate *responsibility* and newer, more progressive models based on corporate *accountability*.[29] Given the increasing rise of non-governmental market initiatives there are also calls that NGO programmes should be accredited to ensure that they too have adequate governance systems to address problems of accountability, transparency and internal democracy. This has led to several national and international level projects aimed at NGO certification. The debate is now whether these should follow the lines of self-regulation or governmental legislation.[30]

The important point is that there appears to be an intersection between the increased recognition of the need for responsible business practice, and the rise of NSMD governance systems, which, it is argued, are specifically designed to regulate the social domain and reconfigure markets. In NSMD, companies voluntarily choose to impose burdens on themselves in order to meet behavioural norms at the global level, most notably the expectation enshrined in the Rio Declaration that sustainable development requires increased public participation. NSMD constitutes a quest for legitimacy at two levels: internally, governance systems need to configure

themselves in a way that means that their core constituents will
accept their legitimacy; and externally, to meet global norms. This
then allows them to actively shape normative behavioural expectations
in directions that best suit themselves and their constituents. Consequently, there is an active contestation between different systems to 'tip
the scales' in favour of certain norms over others.[31] On this view, it is
fair to say that participation, accountability, transparency, democracy
of decision-making, and the other governance arrangements that have
been examined in this study, are at the centre of this contestation. That
such issues should be central to discussions about the role of the private sector in global society 'turns our notions of world politics upside
down. States are sovereign and supposed to be in charge.'[32] However, it
should be reiterated that the success of such private systems of authority to solve environmental problems is not certain. There is a danger
that the competition between systems in one sector (as is the clearly the
case with forest certification schemes) can fragment 'good' behaviour
in the market, and create a market advantage for unsustainable companies. This leads to the conclusion that 'effective institutional design is
essential'.[33] This would appear to confirm the conclusion that being able
to evaluate quality of governance is vital to both improving institutional
performance and design. It also makes the need for global standards all
the more pressing.

Appendix

Key informants

Table A.1 Table of key informants

Name	Organisation or profession at time of interview	Place of interview	National context	Group
Luis Astorga[1]	Association of Forest Engineers for Native Forest	Bonn, Germany	Chile	NGO
Tony Bartlett	General Manager, Forest Industries Branch, Department of Agriculture, Fisheries and Forestry	Canberra, Australia	Australia	Other
Alexander Buck	Deputy Executive Director, International Union of Forest Research Organisations	Vienna, Austria	International	Other
Franz Fiala	Consumer Council, Austrian Standards Institute (ANEC)	Vienna, Austria	Austria	NGO
Graham Drake	Head of Conformity Assessment, ISO[2]	Sydney, Australia	International	Other
Bernard de Galembert	Director, Confederation of European Paper Industries[3]	Brussels, Belgium	EU	Business
James Griffitths	Director, Sustainable Forest Products Industry Sustaining	Geneva, Switzerland	International	Business

---

[1] At the time of interview, a member of the FSC (social chamber, South).
[2] At the time of interview, no longer working for ISO, having recently returned to Australia.
[3] CEPI is an extraordinary member of PEFC.

	Ecosystems Initiative, World Business Council for Sustainable Development			
Nils Hager[4]	Manager, Forest Management and Certification Forests for Life Programme WWF International	Gland, Switzerland	International	NGO
Sini Harkki[5]	Forest Campaigner, Finnish Association for Nature Conservation	Joensuu, Finland	Finland	NGO
John Henry	Secretary, Sub-Committee on Environmental Labelling, TC 207; Director, International and Standardisation Policy, Standards Australia	Sydney, Australia	International; Australia	Other
Esa Härmälä	President, Central Organisation of Finnish Farmers and Private Forest Owners (MTK); President, Confederation of European Private Forest Owners	Telephone interview	Finland; EU	Business
Pierre Hauselmann[6]	Director, Pi Environmental Consulting	Lausanne, Switzerland	International	Business

---

[4] At the time of interview, the representative of WWF International in the FSC (environmental chamber, North).

[5] Co-author, with Matti Liimatainen, of *Anything Goes?* (Helsinki: Greenpeace Nordic and the Finnish Nature League, 2001).

[6] Co-author with Nancy Vallejo of *PEFC: An Analysis* (Pully: Pi Environmental Consulting, 2001) and *Governance and Multi-stakeholder Processes* (Winnipeg: International Institute for Sustainable Development, 2004); author *ISO Inside Out: ISO and Environmental Management* (Godalming: WWF International, 1997), FSC member (economic chamber, North).

Table A.1 (Continued)

Name	Organisation or profession at time of interview	Place of interview	National context	Group
Dick Hortensius[7]	Senior Standardisation Consultant Management Systems, Nederlands Normalisatie-Instituut (NEN)	Delft, Netherlands	Netherlands	Other
Matti Ikonen[8]	Research Assistant – UNEP project, Department of Law, University of Joenssu	Joensuu, Netherlands	Finland	NGO
Outi Jääskö	Member of the Inari Reindeer Herding Cooperative	Inari, Finland	Finland	NGO
Dusan Jovic	Senior adviser, Ministry of Agriculture, Forestry and Water Management-Directorate of Forests-Group for Forest Policy and International Cooperation	Joensuu, Finland	Serbia	Other
Auvo Kaivola	Secretary, Finnish Forest Certification Council	Helsinki, Finland	Finland	Other
Dag Karlsson	District representative, Forest Workers and Woodworkers Union	Bonn, Germany	Sweden	NGO
Jutta Kill[9]	Campaigner, Climate Change and Forest Issues, FERN	Moreton-in-Marsh, United Kingdoms	EU	NGO

---

[7] Author of 'ISO 1400 and Forestry Management: ISO Develops "Bridging Document"' *ISO 9000 + ISO 14000 News* 4 (1999), pp. 11–20.

[8] Previously, forest campaigner, Finnish Nature League, and Greenpeace Finland.

[9] At the time of interview, the representative of Fern in the FSC (environmental chamber, North).

Leontien Krul	FERN	Brussels, Belgium	EU	NGO
Heiko Liedeker	Forest Stewardship Council	Bonn, Germany	N/A	Other
Anders Lindhe[10]	Private consultant	Joensuu, Finland	Sweden	NGO
Andrei Laletin	Friends of the Siberian Forest	Cambridge, United Kingdom	Russia	NGO
Olli Maininnen	Chair, Finnish Nature League (FNL) Forest Group	Helsinki, Finland	Finland	NGO
Rob McLagan[11]	Chief Executive, New Zealand Forest Owners' Association	Bonn, Germany	New Zealand	Business
Saskia Ozinga[12]	Fern	Moreton-in-Marsh, United Kingdom	EU	NGO
Jari Parviainen[13]	Director, Finnish Forest Research Institute, Joensuu Research Centre	Joensuu, Finland	Finland	Other
Duncan Pollard	Head, European Forest Programme, WWF International	Gland, Switzerland	International	NGO
Carl de Schepper	Head, Forest Policy Unit, Agency for Nature and Forests, Ministry of the Environment, Flemish Government	Brussels, Belgium	Belgium	Other

---

[10] Previously, forest campaigner, WWF Sweden, and certification officer, WWF International.

[11] Retired, one month prior to being interviewed.

[12] Author of *Behind the Logo: An Environmental and Social Assessment of Forest Certification Schemes* (Moreton-in-Marsh: Fern, 2001) and *Footprints in the Forest: Current Practice and Future Challenges in Forest Certification* (Moreton-in-Marsh: Fern, 2004).

[13] Also chair of the Finnish Forest Certification Council, and previously also with the Ministry of Foreign Affairs, Finland.

Table A.1   (Continued)

Name	Organisation or profession at time of interview	Place of interview	National context	Group
Matthias Schwoerer	Head, International Forest Policy Division, Federal Ministry of Food, Agriculture and Consumer Protection	Bonn, Germany	Germany	Other
John Scotcher[14]	Private consultant	Bonn, Germany	South Africa	Business
Markku Simula[15]	Private Consultant	Telephone interview	Finland	Business
Karola Taschner	Scientific adviser, European Environmental Bureau (EEB), Chair, European Environmental Citizens' Organisation for Standardisation (ECOS)	Brussels, Belgium	EU	NGO
Hannu Valtanen[16]	Senior Vice President, Forest Policy, Finnish Forest Industries Federation	Telephone interview	Finland	Business
Tiina Vähänen	Forest Officer, Food and Agriculture Organisation	Rome, Italy	International	Other
Matthew Wenban-Smith	Head, Policy and Standards Unit, Forest Stewardship Council International Centre	Bonn, Germany	International	Other
Andrew Wilson[17]	Private individual	Canberra, Australia	New Zealand, Australia	Other

---

[14] Previously, environmental manager South African Pulp and Paper Industries (SAPPI).

[15] The interviewee described their work as having been associated with consultancy assignments related to certification in general and certification of sustainable forest management.

[16] Previously, first vice chairperson, PEFC Council.

[17] This interviewee was interviewed in a personal capacity, but has worked for forestry-related government agencies in Australia and New Zealand.

Notes

1 Introduction

1. United Nations, *Agenda 21: Programme of Action for Sustainable Development, Rio Declaration on Environment and Development, Statement of Forest Principles* (New York: United Nations Publications Department of Public Information, 1993), p. 10 and pp. 230–235.
2. Sherry Arnstein, 'A Ladder of Citizen Participation', *Journal, American Institute of Planners* 35(4) (1969), pp. 216–224, at p. 216.
3. John G. Ruggie, 'Taking Embedded Liberalism Global: The Corporate Connection', in *Taming Globalisation: Frontiers of Governance*, ed. David Held and Matthias Koenig-Archibugi (Cambridge: Polity Press, 2003), pp. 93–129, at p. 93.
4. David Sonnenfeld and Arthur Mol's 'Globalization and the Transformation of Environmental Governance: An Introduction', *American Behavioral Scientist* 45(9) (2002), pp. 1318–1339, at p. 1323.
5. James Rosenau, 'Change, Complexity and Governance in a Globalising Space', *Debating Governance: Authority, Steering and Democracy*, ed. Jon Pierre (Oxford and New York: Oxford University Press, 2000), pp. 167–200, at pp. 167–168.
6. Jan Kooiman, 'Social-Political Governance: Introduction', in *Modern Governance: New Government Society Interactions*, ed. Jan Kooiman (London: Sage, 1993), pp. 1–8, at p. 3.
7. Jan Kooiman, 'Findings, Speculations and Recommendations', in *Modern Governance*, ed. Jan Kooiman (London: Sage, 1993), pp. 249–262, at pp. 250–251.
8. Martijn Van Vliet, 'Environmental Regulation of Business: Options and Constraints for Communicative Governance', in *Modern Governance*, ed. Jan Kooiman (London: Sage Publications, 1993), pp. 105–118, at pp. 108–109.
9. Jürgen Habermas, *Between Facts and Norms: Contributions to a Discourse Theory of Law and Democracy* (Oxford: Blackwell, 1996), pp. 108–111 and pp. 458–459.
10. Rod Rhodes, *Understanding Governance: Policy Networks, Governance, Reflexivity and Accountability* (Buckingham: Open University Press, 1997), p. 48; Lester Salomon, 'The New Governance and the Tools of Public Action: An Introduction', in *The Tools of Government: A Guide to the New Governance*, ed. Lester Saloman (Oxford: Oxford University Press, 2002), pp. 1–41.
11. Jan Kooiman, 'Societal Governance: Levels, Models, and Orders of Social-Political Interaction', in *Debating Governance: Authority, Steering and Democracy*, ed. Jon Pierre (Oxford: Oxford University Press, 2000), p. 163.
12. Jon Pierre and B. Guy Peters, *Governance, Politics and the State* (London: Macmillan, 2000), p. 14.
13. Kooiman, 'Findings, Speculations and Recommendations', p. 259.

14. Michael Zürn and Mathias Koenig-Archibugi, 'Conclusion II: Modes and Dynamics of Global Governance', in *New Modes of Governance in the International System: Exploring Publicness, Delegation and Inclusion*, ed. Mathias Koenig-Archibugi and Michael Zürn (London: Palgrave Macmillan, 2006), pp. 236–254, at p. 251; see also Gerry Stoker, 'The Challenge of Urban Government', in *Debating Governance: Authority, Steering and Democracy*, ed. Jon Pierre (Oxford: Oxford University Press, 2000), p. 107.

15. Anne Mette Kjaer, *Governance* (Cambridge: Polity Press, 2004), p. 12, citing Fritz W. Scharpf, *Games Real Actors Play. Actor-Centered Institutionalism in Policy Research* (Boulder, CO: Westview Press, 1997), p. 153.

16. Steven Bernstein and Benjamin Cashore, 'Nonstate Global Governance: Is Forest Certification a Legitimate Alternative to a Global Forest Convention?', in *Hard Choices, Soft Law: Combining Trade, Environment, and Social Cohesion in Global Governance*, eds John H. Kirton and M. J. Trebilcock (Aldershot: Ashgate, 2004), pp. 33–64, p. 41.

17. Kooiman, 'Societal Governance', p. 159.

18. Bas Arts, 'Non-state Actors in Global Governance: New Arrangements Beyond the State', in *New Modes of Governance in the Global System*, ed. Mathias Koenig-Archibugi and Michael Zürn (Basingstoke: Palgrave Macmillan, 2006), pp. 177–200, at p. 178.

19. Patricia Birnie, 'The UN and the Environment', in *United Nations, Divided World: The UN's Roles in International Relations*, ed. Adam Roberts and Benedict Kingsbury (Oxford: Oxford University Press, 2000), pp. 327–383, at p. 372.

20. Robert Falkner, 'Private Environmental Governance and International Relations: Exploring the Links', *Global Environmental Politics* 3(2) (2003), pp. 72–87, at pp. 72–73.

21. Virginia Haufler, *A Public Role for the Private Sector: Industry Self-regulation* (Washington: Carnegie Endowment for International Peace, 2001), p. 1.

22. Ibid., p. 121.

23. Andrew Jordan, Rüdiger K. W. Wurzel and Anthony Zito, 'The Rise of "New" Policy Instruments in Comparative Perspectives: Has Governance Eclipsed Government?', *Political Studies* 53 (2005), pp. 441–469, at pp. 480–494.

24. Sonnenfeld and Mol, 'Globalization and the Transformation of Environmental Governance', pp. 1318–1323.

25. Ruggie, 'Taking Embedded Liberalism Global', pp. 104–109.

26. Rod Rhodes, *Understanding Governance*, p. 48.

27. Peters, 'Governance and Comparative Politics', pp. 50–51.

28. Mathias Koenig-Archibugi, 'Introduction: Institutional Diversity in Global Gvernance', in *New Modes of Governance in the Global System*, ed. Mathias Koenig-Archibugi and Michael Zürn (Basingstoke: Palgrave Macmillan, 2006), pp. 1–30, at Footnote 13, p. 24.

29. Koenig-Archibugi, 'Introduction: Institutional Diversity in Global Governance', pp. 14–15.

30. Iris Marion Young, *Inclusion and Democracy* (Oxford: Oxford University Press, 2000), pp. 8–13.

31. Michael Warren, 'What Can Democratic Participation Mean Today?', *Political Theory* 30(5) (2002), pp. 677–701, at p. 695.

32. Gerald Berger, 'Reflections on Governance: Power Relations and Policy Making in Regional Sustainable Development', *Journal of Environmental Policy and Planning* 5(3) (2003), pp. 219–234, at pp. 224–225.
33. Katharine Farrell, 'Recapturing Fugitive Power: Epistemology, Complexity and Democracy', *Local Environment* 9(5) (2003), pp. 469–470, p. 476; Stijn Smismans, *Law, Legitimacy, and European Governance* (Oxford: Oxford University Press, 2004), p. 26.
34. Jan Scholte, 'Civil Society and Democratically Accountable Global Governance', *Government and Opposition* 39(2) (2004), pp. 211–233, at pp. 223–225.
35. Michael Mason, *Environmental Democracy* (New York: St Martin's Press, 1999), pp. 72–73.
36. David Held, Anthony G. McGrew, David Goldblatt and Jonathan Perraton, *Global Transformations: Politics, Economics and Culture* (London: Polity Press, 1999), p. 447.
37. Rosenau, 'Change, Complexity and Governance in a Globalising Space', p. 193.
38. Robert Keohane, 'Global Governance and Democratic Accountability', in *Taming Globalization: Frontiers of Governance*, ed. David Held and Mathias Koenig-Archibugi (Cambridge: Polity Press, 2003), pp. 130–159, at p. 137.
39. Pierre and Peters, *Governance, Politics and the State*, pp. 195–196.
40. Joseph Stiglitz, 'Globalization and Development', in *Taming Globalisation*, ed. David Held and Mathias Koenig-Archibugi (Cambridge: Polity Press, 2003), pp. 47–67, at p. 63.
41. Dieter Kerwer, 'Governing Financial Markets by International Standards', in *New Modes of Governance in the Global System*, ed. Mathias Koenig-Archibugi and Michael Zürn (Basingstoke: Palgrave Macmillan, 2006), pp. 77–100, at p. 83.
42. Smismans, *Law, Legitimacy, and European Governance*, p. 22.
43. Scholte, 'Civil Society and Democratically Accountable Global Governance', pp. 211–233.
44. Berger, 'Reflections on Governance', pp. 224–225.
45. Keohane, 'Global Governance and Democratic Accountability', p. 139.
46. David Held, 'Executive to Cosmopolitan Multilateralism', in *Taming Globalization*, ed. David Held and Mathias Koenig-Archibugi (Cambridge: Polity Press, 2003), pp. 174–177.
47. Lawrence Susskind, *Environmental Diplomacy: Negotiating More Effective Global Agreements* (New York, Oxford: Oxford University Press, 2004), p. 7.
48. Jørgen Wettestad, 'Designing Effective Environmental Regimes: The Conditional Keys', *Global Governance* 7(3) (2001), pp. 317–341, at pp. 318–331.
49. James E. Crowfoot and Julia M. Wondolleck, *Environmental Disputes: Community Involvement in Conflict Resolution* (Washington, DC and Clovello: Island Press, 1990), p. 261.
50. Cary Coglianese, *Is Consensus an Appropriate Basis for Regulatory Policy?* (Cambridge, MA: Harvard University Press, 2000), pp. 4–6.
51. Van Vliet, 'Environmental Regulation of Business', pp. 107–108.
52. John Dryzek, *The Politics of the Earth: Environmental Discourses* (New York: Oxford University Press, 1990), p. 200.
53. Stoker, 'The Challenge of Urban Government', pp. 91–109, at p. 105.
54. Pierre and Peters, *Governance, Politics and the State*, p. 31.

55. Durwood Zaelke, Donald Kainaru and Eva Kružíková, *Making Law Work*, Vol. 1 (London: Cameron May, 2005), p. 22, citing various sources.
56. Jon Birger Skjærseth, Olav Schram Stokke and Jørgen Wettestad, 'Soft Law, Hard Law, and Effective Implementation', *Global Environmental Politics* 6(3) (2006), p. 105, footnote 11, following Arild Underdal, 'The Concept of Regime Effectiveness', *Cooperation and Conflict* 27 (1992), pp. 227–240.
57. Oran R. Young, 'Hitting the Mark: Why are Some Environmental Agreements More Effective Than Others?', *Environment* 20 (1999), reproduced in *Making Law Work: Environmental Compliance & Sustainable Development*, ed. Durwood Zaelke, Donald Kainaru and Eva Kružíková, Vol. 1 (London: Cameron May, 2005), p. 189.
58. Carl Folke, Thomas Hahn, Per Olsson and Jon Norberg, 'Adaptive Governance of Social-Ecological Systems', *Annual Review of Environment and Resources* 30 (2005), pp. 463–464.
59. Gulbrandsen, 'Sustainable Forestry in Sweden', p. 253.
60. This approach is adapted from Erik M. Lammerts van Beuren and Esther M. Blom, *Hierarchical Framework for the Formulation of Sustainable Forest Management Standards* (Leiden: The Tropenbos Foundation, 1997), pp. 5–34.
61. Montréal Process, *Criteria and Indicators for the Conservation and Sustainable Management of Temperate and Boreal Forest Ecosystems*, 2nd edn (n.p.: Montréal Process, 1999), p. 5.
62. Lammerts van Beuren and Blom, *Hierarchical Framework*, pp. 20–34.
63. Kooiman, 'Findings, Speculations and Recommendations', p. 260.
64. This term first appears in the United Nations Declaration on the Right to Development, adopted by General Assembly resolution 41/128 of 4 December 1986 (John Gaventa, 'Making Rights Real: Exploring Citizenship, Participation and Accountability', *Journal of International Development Studies* 33(2) (2002), pp. 1–11).
65. See for example John Dryzek and Valerie Braithwaite, 'On the Prospects for Democratic Deliberation: Values Analysis Applied to Australian Politics', *Political Psychology* 21(2) (2000), pp. 241–266.
66. Oran R. Young and Marc A. Levy, 'The Effectiveness of International Environmental Regimes', in *The Effectiveness of International Environmental Regimes: Causal Connections and Behavioural Mechanisms*, ed. Oran R. Young (Cambridge, MA: MIT Press, 1999), pp. 3–4.
67. Lammerts van Beuren and Blom, *Hierarchical Framework*, p. 35.
68. Ibid., p. 34, citing H. G. Baharuddin and M. Simula, 'Timber Certification in Transition. Study on the Development in the Formulation and Implementation of Certification Schemes for all Internationally-Related Timber and Timber Products', report prepared for the International Tropical Timber Organisation, 1996.
69. David Humphreys, *Forest Politics: The Evolution of International Cooperation* (London: Earthscan, 1996), pp. 2–15.
70. Acknowledgment goes once again to Dr Benjamin Cashore for pointing this out.
71. Andrew D. Leslie, 'The Impacts and Mechanics of Certification', *International Forestry Review* 6(1) (2004), pp. 30–39, at p. 34.
72. Bernstein and Cashore, 'Nonstate Global Governance', pp. 33–64, p. 37; Benjamin Cashore, Graeme Auld and Deanna Newsom, *Governing Through*

Markets: Forest Certification and the Emergence of Non-State Authority (New Haven and London: Yale University Press, 2004), Table 1.6 and p. 27. They also include the demand of products by 'purchasers further down the supply chain' as a condition of market-driven governance (ibid.).

73. Arts, 'Non-state Actors in Global Environmental Governance', p. 194.
74. Phillip H. Pattberg, 'The Forest Stewardship Council: Risk and Potential of Private Forest Governance', *Journal of Environment and Development* 14(3) (2005), p. 365, following T. A. Schwandt, *Qualitative Inquiry: A Dictionary of Terms* (Thousand Oaks, CA: Sage, 1997).

2 Governance and Forest Management

1. Arts, 'Non-state Actors in Global Environmental Governance', p. 178.
2. Norah A. Mackendrick, 'The Role of the State in Voluntary Environmental Reform: A Case Study of Public Land', *Policy Sciences* 38 (2005), pp. 21–44, at p. 22.
3. Christine Overdevest, 'Codes of Conduct and Standard Setting in the Forest Sector: Constructing Markets for Democracy?', *Relations Industrielles/Industrial Relations* 59(1) (2004), pp. 172–197, at p. 192.
4. WCED, *Our Common Future* (Melbourne: Oxford University Press, 1987), p. ix.
5. Patricia Birnie, 'The UN and the Environment', pp. 366–368.
6. Ibid.
7. David Humphreys, 'Redefining the Issues: NGO Influence on International Forest Negotiations', *Global Environmental Politics* 4(2) (2004), pp. 51–74, at p. 60.
8. Brook Boyer, 'Multilateral Negotiation Simulation Exercise: The Sustainable Management and Conservation of Forests', in *International Environmental Law-making and Diplomacy Review 2005*, ed. Marko Berglund (Joensuu: University of Joensuu Press, 2005), pp. 299–310, at p. 304.
9. Humphreys, *Forest Politics*, pp. 83–88.
10. Margaret E. Keck and Kathryn Sikkink, *Activists Beyond Borders: Advocacy Networks in International Politics* (New York: Cornell University Press, 1998), pp. 154–156.
11. Humphreys, *Forest Politics*, pp. 83–88 citing UN Document, 'Non-Legally Binding Authoritative Statement of Principles for a Global Consensus on the Management, Conservation and Sustainable Development of All Types of Forests', A/CONF.151/26 (Vol. III).
12. Department of Economic and Social Affairs/Secretariat of the United Nations Forum on Forests, 'United Nation Forum on Forests. Global Partnership: For Forests For People', Fact Sheet 1 (2004), http://www.un.org/esa/forests/factsheet.pdf, accessed 28 October 2004.
13. Michael Howlett and Jeremy Rayner, 'Globalization and Governance Capacity: Explaining Divergence in National Forest Programs as Instances of "Next Generation" Regulation in Canada and Europe', *Governance* 19(2) (2006), pp. 251–275, at p. 261.
14. Bernstein and Cashore, 'Non-state Global Governance', p. 39.

15. Saskia Ozinga, *Behind the Logo: An Environmental and Social Assessment of Forest Certification Schemes* (Moreton-in-Marsh: Fern, 2001), p. 23.
16. Dick Hortensius, 'ISO 14000 and Forestry Management: ISO Develops "Bridging" Document', *ISO 9000-ISO 14000 NEWS* 4 (1999), pp. 11–20, at p. 13. He also lists the International Tropical Timber Organization and the African Timber Organization Initiative.
17. Christopher Elliott, *Forest Certification: A Policy Perspective* (Bogor: Center for International Forestry Research, 2000), pp. 50–51.
18. MCPFE, 'Sustainable Forest Management in Europe, Special Report on the Follow-up on the Implementation of Resolutions H1 and H2 of the Helsinki Ministerial Conference', in *Follow-up Reports on the Ministerial Conferences on the Protection of Forests in Europe*, ed. Liaison Unit in Lisbon, Vol. 2 (Lisbon: Ministry of Agriculture, Rural Development and Fisheries of Portugal, 1998), pp. 6, 258–259.
19. Commonwealth of Australia, *Assessing the Sustainability of Forest Management in Australia* (Canberra: Department of Primary Industries and Energy, Forests Division and Montreal Implementation Group, undated), p. 1.
20. Commonwealth of Australia, *Australia's First Approximation Report For The Montreal Process* (Canberra: Montreal Implementation Group, 1997), p. v (emphasis added).
21. Commonwealth of Australia, 'Assessing the Sustainability of Forest Management', p. 1.
22. http://www.un.org/esa/forests/faq.html, accessed 15 March 2007.
23. Saskia Ozinga, *Footprints in the Forest: Current Practice and Future Challenges in Forest Certification* (Moreton-in-Marsh: FERN, 2004), p. 15, quoting from ECOSOC document E/CN.17/1997/12, 'Report of the Ad Hoc Intergovernmental Panel of Forests on its Fourth Session', 20 March 1997.
24. Howlett and Rayner, 'Globalization and Governance Capacity', pp. 255–256.
25. Peter Glück, Jeremy Rayner, Benjamin Cashore, 'Changes in the Governance of Forest Resources', in *Forests in the Global Balance*, ed. G. Mery, R. Alfaro, M. Kaninnen, and M. Lobovikov (Helsinki: IUFRO, 2005), pp. 51–74, at p. 58.
26. David Humphreys, *Logjam: Deforestation and the Crisis of Global Governance* (London: Earthscan, 2006), p. 215.
27. Humphreys, *Forest Politics*, pp. 83–88.
28. Keck and Sikkink, *Activists Beyond Borders*, p. 160.
29. Fred Gale, *The Tropical Timber Trade Regime* (Basingstoke: Macmillan Press, 1998), pp. 159–161.
30. Keck and Sikkink, *Activists Beyond Borders*, p. 153.
31. Humphreys, *Forest Politics*, pp. 55–57. See also Humphreys, *Logjam*, p. 162. A new Agreement was negotiated in 2006 (ibid.).
32. Humphreys, *Forest Politics*, pp. 33–34, 167.
33. Chris Tollefson, Fred Gale and David Haley, *Setting the Standard: Certification, Governance and the Forest Stewardship Council* (Vancouver: UBC Press, 2008), p. 25.
34. Simon Counsell and Loraas, *Trading in Credibility: The Myth and Reality of the Forest Stewardship Council* (London and Oslo: Rainforest Foundation, 2002), p. 11.
35. Elliott, *Forest Certification*, 2000, p. 1.

36. Counsell and Loraas, *Trading in Credibility*, p. 11.
37. Humphreys, *Forest Politics*, pp. 66–74.
38. Counsell and Loraas, *Trading in Credibility*, p. 11.
39. Humphreys, *Logjam*, p. 116.
40. Humphreys, *Forest Politics*, p. 72, citing ITTO document PCM(V)/D.1, 'Report to the International Tropical Timber Council, Fifth Session of the Permanent Committee on Economic Information and Market intelligence', 03 November 1989.
41. Elliott, *Forest Certification*, p. 46.
42. Counsell and Loraas, *Trading in Credibility*, p. 11; Humphreys, *Forest Politics*, p. 72.
43. David Humphreys, 'The Certification Wars: Forest Certification Schemes as Sites for Trade-environment Conflicts', paper presented to the Privatizing Environmental Governance panel 46th annual convention of the International Studies Association Honolulu, Hawaii 1–5 March 2005, 45 pp., at p. 4.
44. Humphreys, *Forest Politics*, pp. 72–74.
45. Kate Heaton, 'Smart Wood Program', in *International Conference on Certification and Labelling of Products from Sustainably Managed Forests*, ed. Lorraine Cairnes (Canberra: Australian Government Publishing Service, 1996), p. 69.
46. Elliott, *Forest Certification*, pp. 1–74.
47. Arts, 'Non-state Actors in Global Environmental Governance', p. 193; Counsell and Loraas, *Trading in Credibility*, p. 13.
48. Elliott, *Forest Certification*, p. 48, citing Christopher Elliott and Francis Sullivan, *Incentives and sustainability: Where is ITTO Going?* (Gland: WWF International, 1991), pp. 5–6.
49. Humphreys, *Forest Politics*, pp. 66–74.
50. Fred Gale and Cheri Burda, 'The Pitfalls and Potential of Eco-certification as a Market Incentive for Sustainable Forest Management', in *The Wealth of Forests: Markets, Regulation, and Sustainable Forestry*, ed. Chris Tollefson (Vancouver: University of British Columbia Press, 1997), pp. 278–296, at p. 280.
51. Hortensius, 'ISO 14000 and Forestry Management', p. 13.
52. Elliott, *Forest Certification*, p. 68; Counsell and Loraas, *Trading in Credibility*, p. 12.
53. Elliott, *Forest Certification*, p. 68, following Christopher Upton and Stephen Bass, *The Forest Certification Handbook* (London: Earthscan, 1995), p. 148.
54. Counsell and Loraas, *Trading in Credibility*, p. 13.
55. Bernstein and Cashore, 'Nonstate Global Governance', p. 38.
56. Counsell and Loraas, *Trading in Credibility*, pp. 12–13.
57. Jennifer Clapp, 'Global Environmental Governance for Corporate Responsibility and Accountability', *Global Environmental Politics*, 5(3) (2005), pp. 23–34, at p. 25.
58. Matthew Potoski and Aseem Prakash, 'Regulatory Convergence in Nongovernmental Regimes? Cross-National Adoption of ISO 14001 Certifications', *The Journal of Politics* 66(3) (2004), p. 888, citing Michael E. Porter, 'America's Green Strategy', *Scientific American* 264(4) (1991), p. 168.
59. Arts, 'Non-state Actors in Global Governance', p. 190.

60. Kollman and Prakash, 'EMS-based Environmental Regimes as Club Goods', p. 50.
61. Saeed Parto, 'Aiming Low', in *Voluntary Initiatives: The New Politics of Corporate Greening*, ed. Robert B. Gibson (Peterborough: Broadview Press, 1999), pp. 182–198, at p. 182 (footnote 4).
62. Jennifer Clapp, 'Standard Inequities', in *Voluntary Initiatives*, ed. Robert B. Gibson (Peterborough: Broadview Press, 1999), pp. 199–210, at p. 201.
63. Parto, 'Aiming Low', p. 183.
64. Hortensius, 'ISO 14000 and Forestry Management', p. 14.
65. Susan Summers Raines, 'Perceptions of Legitimacy and Efficacy in International Environmental Management Standards: The Impact of the Participation Gap', *Global Environmental Politics* 3(3) (2003), pp. 47–78, at p. 50.
66. Kollman and Prakash, 'EMS-based Environmental Regimes as Club Goods', footnote 13, p. 63.
67. Morikawa and Morrison, *Who Develops ISO Standards?* p. 6.
68. Jennifer Clapp, 'Standard Inequities', p. 201.
69. Saeed Parto, 'Aiming Low', p. 191.
70. Humphreys, *Logjam*, p. 127.
71. Hannes Mäntyranta, *Forest Certification: An Ideal that Became an Absolute*, trans. Heli Mäntyranta (Helsinki: Metsälehti Kustannus, 2002), pp. 47–154.
72. Humphreys, *Logjam*, p. 20.
73. Ibid.
74. Glück et al., 'Changes in the Governance of Forest Resources', p. 55.
75. David Humphreys, *Logjam*, p. 190.
76. Erik Hysing and Jan Olsson, 'Sustainability Through good Advice? Assessing the Governance of Swedish Forest Biodiversity', *Environmental Politics* 14(4) (2005), pp. 510–526, 514–523.
77. Glück et al., 'Changes in the Governance of Forest Resources', pp. 60–61.
78. Benjamin Cashore, Graeme Auld and Deanna Newsom, *Governing Through Markets: Forest Certification and the Emergence of Non-State Authority* (New Haven and London: Yale University Press, 2004), pp. 4–29, 241–247.
79. Errol Meidinger, 'The Administrative Law of Global Private-Public Regulation: The Case of Forestry', *The European Journal of International Law* 17(1) (2006), pp. 47–87, at pp. 30–45.
80. Glück et al., 'Changes in the Governance of Forest Resources', p. 55, citing James G. March and Johan P. Olsen, *Rediscovering Institutions: The Organizational Basis of Politics* (New York: The Free Press, 1989).
81. Elliott, *Forest Certification*, p. 23.
82. Cashore et al., *Governing Through Markets*, pp. 221–241.
83. Lars H. Gulbrandsen, 'Sustainable Forestry in Sweden: The Effect of Competition Among Private Certification Schemes', *The Journal of Environmental Development* 14(3) (2005), pp. 338–355, at p. 349.
84. Cashore et al., *Governing Through Markets*, pp. 34–37, pp. 221–241.
85. Meidinger 'The Administrative Law of Global Private-Public Regulation', pp. 47–87, at p. 49.
86. Howlett and Rayner, 'Globalization and Governance Capacity', p. 257.
87. Meidinger, 'The Administrative Law of Global Private-Public Regulation', p. 70.
88. Gulbrandsen, 'Sustainable Forestry in Sweden', p. 351.

89. Magnus Boström, 'Regulatory Credibility and Authority through Inclusiveness: Standardization Organizations in Cases of Eco-labelling', *Organization* 13(3) (2006), pp. 345–367.
90. Meidinger, 'The Administrative Law of Global Private–Public Regulation', pp. 66–83.
91. Elliott, *Forest Certification*, p. 23.
92. Cashore et al., *Governing Through Markets*, pp. 225–247, at p. 227.
93. Humphreys, 'The Certification Wars', pp. 34–35.
94. Ozinga, *Footprints in the Forest*, p. 17.
95. This issue is discussed by Humphreys, *Logjam*, p. 135, citing Fred Gale, '*Caveat Certificatum*: The Case of Forest Certification', in *Confronting Consumption*, ed. Thomas Princen, Michael Maniates and Ken Conca (Cambridge MA: MIT Press, 2002), p. 288.
96. The author would like to thank Dr Benjamin Cashore for making this pertinent observation.

3 Forest Stewardship Council

1. The meeting included WWF, the Rainforest Alliance and Greenpeace, British timber retailer B&Q, the Ecological Trading Company, and The Woodworkers' Alliance for Rainforest Protection (Cashore et al., Governing Through Markets, pp. 1–11; Hannes Mäntyranta, *Forest Certification – An Ideal That Became an Absolute*, trans. Heli Mäntyranta (Helsinki: Metsälehti Kustannus, 2002), p. 18).
2. Mäntyranta lists the following individuals as playing a significant founding role in FSC at this date: Chris Elliott, Senior Forests Officer with WWF International, based in Gland, Switzerland; Julio César Centeno, Professor of Wood Technology in the University of the Andes, Venezuela; Alan Knight, Environmental Coordinator B&Q Plc., Debbie Hammel, Director, Scientific Certification Systems, US; Dagoberto Irias, Honduras; Dominique Irvine, Program Director, Cultural Survival, US; Andrew Poynter, Director, Woodworkers Alliance for Rainforest Protection and owner of A&M Wood Speciality, Inc., Canada.
3. FSC's P&C for Forest Stewardship currently cover: Compliance with laws and FSC principles; Tenure use rights and responsibilities; Indigenous peoples' rights, Community relations and workers' rights; Benefits from the forest; Environmental impact; Management plan; Monitoring and assessment; Maintenance of high conservation value forests; Plantations. The P&C are currently under review.
4. FSC, 'Document 1.1. Forest Stewardship Council A.C. By-Laws', ratified September 1994; editorial revision, October 1996; revised February 1999, August 2000, November 2002, June 2005, June 2006, paragraph 1, p. 1.
5. This book uses 'forest owners' as a generic term, 'private forest owners' to refer to private forest owners of any size, and 'small forest owners' when seeking to emphasise the small-scale element of the sector.
6. FSC, 'Plantations Review', http://www.old.fsc.org/plantations, accessed 10/07/07.
7. Romeijn, *Green Gold*, pp. 5–8.
8. FSC, Annual Report 2002, p. 8.

9. At the time, the author was an individual member of FSC (environment chamber North), and was elected to participate in the group.
10. Paragraph 28, By-Laws, p. 4.
11. Paragraph 31, By-Laws, p. 5.
12. FSC, 'Accreditation – Accreditation of Certification Bodies', http://www.fsc.org/en/about/accreditation/accred-certbod, accessed 21 June 2007.
13. Meidinger, 'The Administrative Law of Global Private-public Regulation', p. 52.
14. Qualified majority is a term used to mean 'more than simple majority: typically two thirds' (Rod Hague and Martin Harrop, *Comparative Government and Politics: An Introduction*, 7th ed. (Basingstoke: Palgrave Macmillan, 2007), p. 47).
15. Gulbrandsen, 'Sustainable Forestry in Sweden', p. 349.
16. Pattberg, 'The Forest Stewardship Council', p. 365.
17. FSC, 'FSC Social Strategy', p. 9.
18. Paragraph 1.4.3, FSC-STD-20-006, p. 4.
19. Kirsti Thornber, 'Certification: A Discussion of Equity Issues', in *Social and Political Dimensions of Forest Certification*, ed. Errol Meidinger, Christopher Elliott and Gerhard Oesten (Remagen-Oberwinter: Forstbuch, 2003), pp. 63–82.
20. FSC-STD-20-003, p. 1.
21. Mäntyranta, *Forest Certification*, p. 25.
22. Paragraph 12, By-Laws, p. 2.
23. Eero Palmujoki, 'Public–private Governance Patterns and Environmental Sustainability', *Environment, Development and Sustainability* 8 (2006), pp. 1–17, at p. 13.
24. Meidinger, 'The Administrative Law of Global Private-public Regulation', p. 69.
25. Pattberg, 'The Forest Stewardship Council', p. 370.
26. Phillip H. Pattberg, 'What Role for Private Rule-Making in Global Environmental Governance? Analysing the Forest Stewardship Council (FSC)', *International Environmental Agreements* 5(2) (2005), pp. 175–189 at p. 185.
27. Pattberg, 'The Forest Stewardship Council', p. 369.
28. Ibid., pp. 366–367.
29. Eckhard Rehbinder, 'Forest Certification and Environmental Law', in *Social and Political Dimensions of Forest Certification*, ed. Errol Meidinger, Christopher Elliott and Gerhard Oesten (Remagen-Oberwinter: Forstbuch, 2003), pp. 331–351, at p. 345.
30. Gulbrandsen, 'Overlapping Public and Private Governance', p. 89 and Rehbinder, 'Forest Certification and Environmental Law', p. 345.
31. Pattberg, 'The Forest Stewardship Council', p. 366.
32. FSC Standard FSC-STD-20-001, Version 2.1, FSC Standard FSC-STD-20-004, Version 2.2, FSC-STD-20-005, Version 2.1, FSC Standard FSC-STD-20-007 Version 2.1.

4 ISO, TC 207 and the 14000 Series

1. Article 2.1, ISO Statutes and Rules of Procedure, cited in Sebastian Oberthür, Matthias Buck, Sebastian Müller, Stefanie Pfahl, Richard G. Tarasofsky, Jacob

Werksman and Alice Palmer, *Participation of Non-governmental Organisations in International Environmental Governance: Legal Basis and Practical Experience* (Berlin: Ecologic/Centre for International and European Environmental Research, 2002), p. 164.

2. Kollman and Prakash, 'EMS-based Environmental Regimes as Club Goods', p. 49.
3. Raines, 'Perceptions of Legitimacy and Efficacy', p. 49.
4. TC 207, http://www.TC207.org, accessed 10 May 2007.
5. ISO/TC 207, 'Increasing the Effectiveness of NGO Participation in ISO TC 207', N590 Rev/ISO/TC NGO TG N28, 30 June 2003, p. 2.
6. ISO/TC 207, 'Draft: Recommendations for an Improved Balance of Stakeholder Participation in ISO TC 207', ISO/TC 207 CAG N437R1 (April 2007).
7. ISO/TC 207, 'Draft ISO/TC 207 Proposed Process to Identify Stakeholder Balance at ISO/TC 207 Meetings', ISO/TC 207 CAG N439R1 (undated), p. 2.
8. Personal communication, 11 March 2010.
9. ANEC, ECOS, Pacific Institute, 'ISO TC 207 "Environmental Management" Gives NGOs the Cold Shoulder', communiqué (undated), pp. 2 and 7.
10. Personal communication, 11 March 2010.
11. *Articles and News*, 'Environment – ISO/TC 207 Considers Industry's Needs', (August, 1995), p. 3.
12. *Articles and News*, '3rd Meeting of ISO/TC 207 in Oslo, Norway' (August, 1995), p. 7.
13. Pierre Hauselmann, *ISO Inside Out: ISO and Environmental Management*, second edition (Godalming: WWF International, 1997), p. 13.
14. Hauselmann, *ISO Inside Out*, p. 8.
15. http://www.iso.org/iso/en/aboutiso/introduction/index.html, accessed 02 May 2007.
16. ISO, 'ISO's structure' http://www.iso.org/iso/structure, accessed 18/06/08.
17. Arts, 'Non-state Actors in Global Governance', p. 193.
18. Potoski and Prakash, 'Regulatory Convergence in Nongovernmental Regimes', p. 885, Jennifer Clapp, 'The Privatization of Global Environmental Governance: ISO 14000 and the Developing World', *Global Governance* 4(3) (1998), pp. 23–34, at p. 295, Falkner, 'Private Environmental Governance', pp. 76–77.
19. Potoski and Prakash, 'Regulatory Convergence in Nongovernmental Regimes', p. 888, and Glück et al., 'Changes in the Governance of Forest Resources', p. 61 (citing Frtiz Scharpf, *Governing in Europe: Effective and Democratic?* (Oxford: Oxford University Press, 1999)).
20. Mari Morikawa and Jason Morrison, *Who Develops ISO Standards? A Survey of Participation in ISO's International Standards Development Processes* (Oakland: Pacific Institute, 2004), pp. 17–18.
21. Raines, 'Perceptions of Legitimacy and Efficacy', p. 48.
22. Brad Edwards, Jill Gravender, Annette Killmer, Genia Schenke and Mel Willis, *The Effectiveness of ISO 14001 in the United States* (Santa Barbara: University of California, 1999), pp. 153, 164.
23. Morikawa and Morison, *Who Develops ISO Standards?* p. 18.
24. Mike Smith, ISO Central Secretariat, personal communication, 01 November 2007.

25. ISO, *Statutes and Rules of Procedure*, 14th ed. (Geneva: ISO, 2000), Clause 3.8, p. 33.
26. Raines, 'Perceptions of Legitimacy and Efficacy'.
27. Oberthür et al., Participation of Non-governmental Organisations, p. 169.
28. ISO/TC 207, 'Business Plan ISO/TC 207 Environmental Management', N726R0/CAG N 376R3 Version 3 (June, 2005).
29. Edwards et al., *The Effectiveness of ISO 14001*, p. 153 and p. 164.
30. ANEC/EEB, 'ANEC/EEB Position Paper on Environmental Management System Standards' (February 2003), p. 1.
31. ISO/TC 207, 'Increasing the Effectiveness', p. 5.
32. Paragraph 1.7 ISO/IEC Guide 2 (1991), cited in Elliott, *Forest Certification*, p. 18.
33. Hauselmann, *ISO Inside Out*, p. 11.
34. Matthew Potoski and Aseem Prakash 'Green Clubs and Voluntary Governance: ISO 14001 and Firms' Regulatory Compliance', *American Journal of Political Science* 49(2) (2005), pp. 235–248.
35. Raines, Susan Summers, 'Judicious Incentives: International Public Policy Responses to the Globalization of Environmental Management', *Review of Policy Research* 23(2) (2006), pp. 473–490.
36. Saeed Parto, 'Aiming Low', in *Voluntary Initiatives: The New Politics of Corporate Greening*, ed. Robert B. Gibson (Peterborough: Broadview Press, 1999), p. 187.
37. Andrew A. King, Michael J. Lenox, and Ann Terlaak, 'The Strategic Use of Decentralized Institutions: Exploring Certification with the ISO 14001 Management Standard', *Academy of Management Journal* 48(6) (2005), pp. 1091–1106, at p. 1104.
38. Edwards et al., *The Effectiveness of ISO 14001*, p. 99.
39. Susan L. K. Briggs, 'Do Environmental Management Systems Improve Performance?', *Quality Progress* 39(9) (2006), p. 78.
40. Nancy Vallejo and Pierre Hauselmann, *PEFC An Analysis* (Pully: WWF, 2001), p. 13, citing BATE's ISO 14001 Update, 'No Link Found between Management Systems and Performance', *Business and the Environment* 7(1) (2001).
41. ANEC/ECOS, 'Joint ANEC/ECOS Comments on the ISO 14000 Series Review', ANEC-ENV-G-030final (October, 2007).
42. In addition to TC 207, two programmes are of most relevance to this study, ISO's activities in the area of social responsibility (ISO 26000) and Technical Committee 224, since they cover the historical development of non-state participation in ISO's social and environmental standards. TC 224 was created in 2001 to develop water management standards.
43. ISO/TC 207 N590, pp. 7–8 (referring to ISO Directives Clause 2.3.5 and Clause 3.3.3).

5 Programme for the Endorsement of Forest Certification Schemes

1. Mäntyranta, *Forest Certification*, p. 97.
2. Finnish Forest Industries Federation, 'Increased credibility for Finnish Forest Certification Approach through European Cooperation', media release 25

August 1998; 'Industry fully supports the Finnish Forest Certification System (FFCS) – FSC-Labelling of Products Becomes Available, Too?' media release 28 October 1998.

3. Mäntyranta, *Forest Certification*, p. 132.
4. Anonymous, 'Pan European Forest Certification', memorandum, 2 September 1999, http://www.faf.de/paneuro_e.htm, accessed 20/09/1999.
5. Ibid.
6. Mäntyranta, *Forest Certification*, p. 245.
7. PEFC, 'PEFC Council Position Statement on Recent WWF Press Release on Austrian Forest Certification Scheme', media release, 08 May 2001.
8. Gabriele Herzog, 'PEFC Austria – Totally Compliant with Austrian Law – Issues First Logo Licenses', *PEFCC Newsletter* 8 (July 2001), p. 5.
9. Ozinga, *Footprints in the Forest*, p. 51.
10. Ibid.
11. *PEFCC Newsletter*, 'PEFC Austria is on the Road to Success – WWF Accusations "Substantially Wrong"', 7 (July 2001), p. 5.
12. *PEFC News*, 'International ENGO Platform Supporting PEFC Certification', 25 (April 2005), p. 2.
13. *PEFC News*, 'PEFC Open Letter To The World Bank And WWF', 21 (September 2004), p. 1.
14. Ozinga, *Behind the Logo*, p. 23.
15. Ivar Korsbakken and Svein Søgnen, 'Norwegian Forest Owners Warn of Trade Barriers', *PEFC News* 35 (January 2007), p. 11.
16. *PEFC News*, 'International Accreditation Forum', p. 1.
17. http://www.pefc.org/internet/html/about_PEFC/4_1137_498.htm (accessed 14 June 2007).
18. *PEFC News*, 'PEFC: Only European? Based On Merely a Handful of Criteria? Not For Indigenous People? No Social Dialogue ...?', 37 (May 2007), p. 2.
19. Zyen and PEFC, 'PEFC Governance Review May, 2008 Final Report', May 2008.
20. The author was invited to present to the panel (Tim Cadman, 'Evaluating Legitimacy and Quality of Forest Governance', contained in Zyen and PEFC, 'Governance Review', appendix H, pp. 98–104).
21. One World Trust, 'The GAP Framework' http://www.oneworldtrust.org/?display=gapframework (accessed 30/09/08).
22. Zyen and PEFC, 'Governance Review', p. 8.
23. WWF, 'WWF Statement Regarding the PEFC Governance Review and the new PEFC Stakeholder Forum', press release, 09 February 2009, p. 2.
24. PEFC Council, 'PEFC Comments on Recent WWF's 2008 Certification Scheme Assessment and Public Statement on PEFC Stakeholder Forum', 05 March 2009.
25. PEFC, 'Terms and Conditions', Annex 1, Chapter 2, October 2004, p. 4.
26. http://www.pef.org/internet/html/about_pefc.htm, accessed 14 June 2007.
27. PEFC, 'PEFC Technical Document', Chapter 4, p. 6.
28. PEFC, 'Rules for Standards Setting', Annex 2, 29 October 2004, Chapter 3.5.1, p. 3.
29. PEFC, 'PEFC Technical Document', Chapter 5, p. 8.
30. Chapter 3.5.3, PEFC Rules for Standard Setting, p. 3.

31. PEFC, 'Endorsement and Mutual Recognition of National Schemes and their Revision', Chapter 3, p. 2; 'Endorsement and Mutual Recognition of National Schemes and their Revision', Chapter 6, pp. 4.
32. PEFC, 'Certification and Accreditation Procedures', Chapter 5, p. 4.
33. Margaret Shannon, 'What is Meant by Public Participation in Forest Certification Processes? Understanding Forest Certification Within Democratic Governance Institutions', in *Social and Political Dimensions of Forest Certification*, ed. Errol Meidinger, Christopher Elliott and Gerhard Oesten (Remagen-Oberwinter: Forstbuch, 2003), pp. 179–198, at p. 179.
34. Meidinger, 'The Administrative Law of Global Private–Public Regulation', p. 70, Humphreys, *Logjam*, p. 139. Humphreys concludes that it is the FSC that is exerting this 'pull'.
35. PEFC, 'PEFC Council Technical Document', Chapter 3.2, p. 6.
36. Ibid.
37. Article 3.1, PEFC Council Statutes, p. 1.
38. Vallejo and Hauselmann, *PEFC: An Analysis*, p. 6.
39. Ibid.
40. Gulbrandsen, 'Sustainable Forestry in Sweden', pp. 343–347.
41. Ozinga, 'The European NGO Position on the PEFC', presentation on behalf of FERN to the Pan-European Forest Certification (PEFC) Seminar and Workshop, Würzburg, 19–21 April 1999, p. 7.
42. Gulbrandsen, 'Sustainable Forestry in Sweden', pp. 343–347.
43. Ibid., p. 352.
44. Ozinga, *Behind the Logo*, p. 23.
45. Articles 6.7 and 6.8, PEFC Council Statutes, p. 4.
46. Humphreys, 'The Certification Wars', pp. 20–21.
47. Rehbinder, 'Forest Certification and Environmental Law', p. 340.
48. Ozinga, *Behind the Logo*, p. 22, footnote 48.
49. PEFC, 'PEFC Rules for Standard Setting', Chapter 6, p. 5.
50. Email to PEFC Secretariat sent 01 November 2007.
51. Savcor Indufor Oy, Effectiveness and Efficiency of FSC and PEFC Forest Certification on Pilot Areas in Nordic Countries (Helsinki: Federation of Nordic Forest Owners' Organisations, 2005), p. 94.
52. Finnish Forest Industries Federation, 'Finnish Forest Industries Need Raw Material from Certified Forests', media release 07 September 2000.
53. PEFC, 'Endorsement and Mutual Recognition of National Schemes and their Revision', Chapter 6, pp. 4–5.
54. PEFC, 'PEFC Technical Document', Chapter 10, p. 13.
55. Yrjö-Koskinen et al., *Certifying Extinction*, p. 9.
56. PEFC, 'PEFC Council Procedures for the Investigation and Resolution of Complaints and Appeals', June 2007, p. 5.
57. Gulbrandsen, 'The Effectiveness of Non-State Governance Schemes', p. 138.
58. Ozinga, *Footprints in the Forest*, p. 15, quoting from ECOSOC document E/CN.17/1997/12, 'Report of the Ad Hoc Intergovernmental Panel of Forests on its Fourth Session', 20 March 1997.
59. Humphreys, 'The Certification Wars', p. 20.
60. Ozinga, *Behind the Logo*, p. 18.

61. Ibid., referring to Sweden. See also Yrjö-Koskinen et al., *Certifying Extinction*, pp. 11–19 (referring to Finland).
62. Vallejo and Hauselmann, *PEFC: An Analysis*, p. 8, referring to PEFC certification in Europe.
63. Benjamin Cashore, Graeme Auld and Deanna Newsom, 'Forest Certification (Eco-Labelling) Programs and Their Policy-making Authority: Explaining Divergence Among North American and European Case Studies', *Forest Policy and Economics* 5 (2003), pp. 225–247, at p. 227; Gulbrandsen, 'Sustainable Forestry in Sweden', pp. 343–352.
64. Humphreys, 'The Certification Wars', pp. 20–21.
65. Ozinga, *Behind the Logo*, p. 18.

6   United Nations Forum on Forests

1. ECOSOC Resolution 2000/35, 'Report of the Fourth Session of the Inter-governmental Forum on Forests', 18 October 2000, http://www.un.org/documents/ecosoc/dec/2000/edec2000-inf2-add3.pdf, accessed 15 March 2005.
2. ECOSOC Resolution 2000/35, Article 1, p. 64.
3. ECOSOC Resolution 2000/35, Article 3 (c), p. 65.
4. UN Document E/CN.18/2001/2, 'Eight-Country Initiative, Shaping the Programme of Work for the United Nations Forum on Forests (UNFF)', 7 February 2001, pp. 3–6.
5. Humphreys, 'Redefining the Issues', p. 68.
6. Radoslav S. Dimitrov, 'Hostage to Norms: States, Institutions and Global Forest Politics', *Global Environmental Politics* 5(4) (2005), pp. 1–24, at p. 9.
7. Humphreys, 'Redefining the Issues', pp. 66–69.
8. UN Document E/2002/42-E/CN.18/2002/42, p. 7.
9. Ibid., p. 39.
10. UN Document E/2001/42/Rev.1-E/CN.18/2001/3/Rev.1, resolution 1/1 para. 33, p. 8.
11. The groups are officially listed for UNFF-3 as: women; youth and children; workers and trade unions; the scientific and technical community; indigenous peoples' organisations; environmental non-governmental organisations; local authorities; small forest owners; and business and industry (UN Document E/2003/42-E/CN.18/2003/13, p. 32).
12. UN Document E/2003/42-E/CN.18/2003/13, pp. 32–36.
13. UN Document E/2004/42-E/CN.18/2004/17, resolution 4/4, para. 6, p. 9.
14. Humphreys, *Logjam*, pp. 103–108.
15. UN Document E/2004/42 E/CN.18/2004/17, decision 4/2, p. 11.
16. UN Document E/2004/42-E/CN.18/2004/17, resolution 4/3, para. 1, p. 6.
17. *Earth Negotiations Bulletin*, 'Polite Listening: Engaging Civil Society', 13(116) (2004), pp. 9–10, at p. 11.
18. UN Document E/2004/42-E/CN.18/2004/17, para. 10, p. 18.
19. Ibid., para. 12, p. 18.
20. Emily Caruso and Leontien Krul, 'Special FERN-FPP Report: UNFF Failing its Mandate – 4th Session of the United Nations Forum on Forests', *EU Forest Watch* (June 2004), pp. 1–3, at p. 3.

21. Caruso and Krul, 'Special FERN-FPP Report', p. 1.
22. *Earth Negotiations Bulletin*, 'Quo Vadis: Contemplating the Post-UNFF Era', 13(116), p. 11.
23. Leontien Krul, Tom Griffiths and Saskia Ozinga, 'Live or Let Die? An Evaluation of the Fifth Session of the United Nations Forum on Forests', *FERN Special Report*, July 2005, p. 3.
24. *Earth Negotiations Bulletin*, 'Multi-stakeholder Dialogue', 13(133) (2005), p. 3.
25. UN, 'Multi-stakeholder Dialogue Note by the Secretariat. Addendum. Discussion Paper Contributed by the Business and Industry Major Group', UN Document E/CN.18/2005/3/Add.1, 24 March 2005, pp. 1–5.
26. UN Document, 'Multi-stakeholder Dialogue Note by the Secretariat. Addendum. Discussion Paper Contributed by the Non-governmental Organizations Major Group', E/CN.18/2005/3/Add.4, 24 March 2005, p. 1.
27. Un document E/CN.18/2005/3/Add.4, p. 4.
28. UN Document E/2006/42-E/CN.18/2006/18, p. 7.
29. Perhaps indicating that: 'UNFF as a purely international process has proved ineffective and that a shift in focus back to the regional level is needed' (Humphreys, *Logjam*, p. 114); 'Member States & Bureau: Country, Organization and Region Led Initiatives'; http://www.un.org/esa/forests/gov.html, accessed 22 July 2010.
30. *Earth Negotiations Bulletin*, 'The Future of the International Arrangement on Forests', 13(144) (2006), p. 12.
31. *Earth Negotiations Bulletin*, 'The Role of UNFF and Other Players', 13(150) (2006), p. 11.
32. Ibid.
33. Dimitrov, 'Hostage to Norms', p. 9.
34. *Earth Negotiations Bulletin*, 'The MPOW: Not Just Another Shopping List', 13(162) (2007), pp. 16–18.
35. *Earth Negotiations Bulletin*, 'The MPOW', p. 19.
36. *Earth Negotiations Bulletin*, 'Annex: MYPOW for 2007–2015', 13(162) (2007), p. 16.
37. Interestingly, SFM ended up in the final NLBI as being referred to as 'a dynamic and evolving concept [which] aims to maintain and enhance the economic, social and environmental values of all types of forests, for the benefit of present and future generations' (UN Resolution 62/98, 'Establishing the Non-Legally Binding Instrument on all Types of Forests', http://daccessdds.un.org/doc/UNDOC/GEN/N07/469/65/PDF/N0746965.pdf?OpenElement, accessed 25 October 2010).
38. *Earth Negotiations Bulletin*, 'The NLBI: Where's The Meat?' 13(162) (2007), p. 18.
39. Ibid.
40. UN Document E/2007/42, p. 44.
41. Ibid.
42. United Nations, 'Discussion Paper Contributed by the Non-governmental Organizations and Indigenous Peoples Major Group', UN Document E/CN.18/2009/13/Add.3 (January 2009), p. 2.
43. *Earth Negotiations Bulletin*, 'Facing the Future', 13(174) (2007), p. 9.
44. http://www.un.org/esa/forests/factsheet.pdf, accessed 28 October 2004.
45. http://www.un.org/esa/forests/factsheet.pdf, accessed 28 October 2004.

46. ECOSOC Resolution 2000/35, Article 1 (b), p. 64.
47. Glück et al., 'Changes in the Governance of Forest Resources', p. 58, citing Peter Glück, 'National Forest Programs – Significance of a Forest Policy Framework', in *Formulation and Implementation of National Forest Programmes – EFI Proceedings No. 30, Volume I: Theoretical Aspects*, ed. Peter Glück, Gerhard Oesten, H. Schanz and K.-R. Volz (Joensuu: European Forestry Institute, 1999), pp. 39–51.
48. 1) formulation and implementation of NFPs; 2) promoting public participation; 3) deforestation and forest degradation; 4) traditional forest-related knowledge; 5) forest-related scientific knowledge; 6) forest health and productivity; 7) criteria and indicators for sustainable forest management; 8) economic, social and cultural aspects of forests; 9) forest conservation and protection of unique types of forests and fragile ecosystems; 10) monitoring, assessment and reporting, and concepts, terminology and definitions; 11) rehabilitation and conservation strategies for forests with low forest cover; 12) rehabilitation and restoration of degraded forest lands, and the promotion of natural and planted forests; 13) maintaining forest cover to meet present and future needs; 14) financial resources; 15) international trade and sustainable forest management; 16) international cooperation in capacity-building, and access to and transfer of environmentally sound technologies to support sustainable forest management.
49. UNFF, 'About UNFF: History and Milestones of International Forest Policy', http://www.un.org/esa/forests/about-history.html, accessed 05 March 2010.
50. Personal communication, 05 March 2010.
51. Jan Kooiman, 'Societal Governance', pp. 151–154.
52. ECOSOC Resolution 2000/35, Article 2 (d), p. 64. PPPs have been identified by Kooiman as one of a range of 'new patterns of interaction between government and society' (Jan Kooiman, 'Social–Political Governance', p. 1).
53. UN Document E/2004/42-E/CN.18/2004/17, para. 25, p. 20.
54. UN Document E/2001/42/Rev.1-E/2001/42/Rev.1, Resolution 1/1 paras 31 and 33 p. 9 (following ECOSOC Resolution 2000/35, Articles 4, (b) and (c)).
55. UN Document E/CN.18/2002/10, p. 2.
56. Department of Economic and Social Affairs/Secretariat of the United Nations Forum on Forests, 'United Nation Forum on Forests. Global Partnership: For Forests For People', Fact Sheet 1 (2004), http://www.un.org/esa/forests/factsheet.pdf, accessed 28 October 2004.
57. Humphreys, *Logjam*, p. 97. It is not clear why Cuba played this role. One interviewee put it down to the fact that Cuba was a 'conservative' country, and often allied itself with other conservatives, such as Saudi Arabia, to prevent NGO interventions (personal interview, 12 September 2006).
58. Bill Mankin, 'MY POW or Yours? Choosing a Leadership Agenda for the UNFF', unpublished paper prepared for CIFOR, January 2007, pp. 17–18.
59. UNFF Secretariat, 'Review of the Effectiveness of the International Arrangement on Forests Analytical Study', undated, p. 11 and Department of Economic and Social Affairs/UNFF Secretariat, 'Implementation of Proposals for Action Agreed by Intergovernmental Panel on Forests and by Intergovernmental Forum on Forests (IPF/IFF) Action for Sustainable Forest Management', December 2005, pp. 37–38.

60. Doris Capistrano, Markku Kanninen, Manuel Guariguata, Chris Barr, Terry Sunderland and David Raitzer, 'Revitalizing the UNFF: Critical Issues and Ways Forward', paper prepared for the Country-led Initiative on Multi-year Programme of Work of the United Nations Forum on Forests: Charting the Way Forward to 2015, Bali, Indonesia, 13–16 February 2007, p. 19.

61. Glück et al., 'Changes in the Governance of Forest Resources', p. 72.

62. UN Document E/2004/42-E/CN.18/2004/17, para. 24, p. 19.

63. UN Document E/2005/42-E/CN.18/2005/18, p. 31.

64. UN, 'United Nations Forum on Forests Bureau of the Sixth Session (UNFF-6 Bureau)' minutes of the fourth meeting, 25–26 January 2006, p. 2.

65. Mankin, 'MY POW or Yours', p. 20. The comments read as follows:

66. Capistrano et al., 'Revitalizing the UNFF', p. 12.

67. UN Document E/2004/42-E/CN.18/2004/17, para. 25, p. 20.

68. Humphreys, *Logjam*, p. 98.

69. UN Document E/2005/42-E/CN.18/2005/18, p. 31.

70. UNFF Resolution 1/1, UN Document E/2001/42/Rev.1-E/2001/42/Rev.1, paras 31 and 33, p. 9 (relating to ECOSOC Resolution 2000/35, Articles 4, (b) and (c)).

71. UNFF Resolution 1/1, UN Document E/2001/42/Rev.1-E/CN.18/2001/3/Rev.1, para. 29, p. 8.

72. UN Document E/2002/42-E/CN.18/2002/42, p. 16.

73. UNFF Secretariat, 'Review of the Effectiveness' 2005, p. 3.

74. UN Document E/2002/42-E/CN.18/2002/42, p. 16.

75. *Earth Negotiations Bulletin*, 'Achieving the Global Objectives and Implementing NLBI', 13(174), p. 7.

76. UNFF secretariat was directly contacted by the researcher via email 01 November 2007 requesting information, and again through its own web page contact facility 06 August 2008. No response was provided on either occasion.

77. ECOSOC Document E/5715/Rev.2, pp. 28–32.

78. ECOSOC Resolution 1993/215, pp. 97–98.

79. ECOSOC Resolution 2000/35 Article 18.

80. Humphreys, *Logjam*, p. 98.

81. *Earth Negotiations Bulletin*, 'A Brief Analysis of UNFF-6', 13(144) (2006), p. 11.

82. Bernstein, Johanna, 'Sustainable Development Governance Challenges in the New Millennium', in *International Environmental Law-making and Diplomacy Review 2004*, ed. Marko Berglund (Joensuu: University of Joensuu Department of Law, 2005), pp. 31–50, at p. 34.

83. Patrick Széll, 'Introduction to the Discussion on Compliance', in *International Environmental Law-making and Diplomacy Review 2004*, ed. Marko Berglund (Joensuu: University of Joensuu Department of Law, 2005), pp. 117–123, at p. 118.

84. Capistrano et al., 'Revitalizing the UNFF', p. 2; Mankin, 'MY POW or Yours', p. 8.

85. Humphreys, *Logjam*, p. 114.

86. Glück et al., *Forests in the Global Balance*, pp. 57–70.

87. Dimitrov, 'Hostage to Norms', pp. 1–24.

88. Ibid., p. 20.
89. Humphreys, *Logjam*, p. 99, emphasis in original.
90. Skjærseth et al., 'Soft Law, Hard Law', p. 119.
91. Steven Bernstein, 'Liberal Environmentalism and Global Environmental Governance', *Global Environmental Politics* 2(3) (2002), pp. 1–16, at p. 12.
92. Mankin, 'MY POW or Yours', pp. 1–28.
93. Ibid., p. 18.
94. Humphreys, *Logjam*, pp. 153 and 166.
95. Mankin, 'MY POW or Yours', p. 3.
96. UNFF Secretariat, 'Implementation of Proposals for Action Agreed by Intergovernmental Panel on Forests and by Intergovernmental Forum on Forests (IPF/IFF): Action for Sustainable Forest Management', http://www.un. org/esa/forests/pdf,publications/proposals-for-action.pdf, accessed 07 March 2010, p. 4.
97. *Earth Negotiations Bulletin*, 'A Brief Analysis', p. 14.
98. UN Document E/2003/42-E/CN.18/2003/13, para. 14, p. 34.

7 Comparative Analysis

1. Joseph E. Stiglitz, 'Globalization and Development', in *Taming Globalisation: Frontiers of Governance*, ed. David Held and Matthias Koenig-Archibugi (Cambridge: Polity Press, 2003), pp. 47–67, at p. 63.
2. Pattberg, 'What Role for Private Rule-Making', p. 182 referring the FSC in particular.
3. Vallejo and Hauselmann, *Governance and Multi-stakeholder Processes*, pp. 5–6.
4. Anne Mette Kjaer, *Governance* (Cambridge and Malden MA: Polity Press, 2004), p. 14.
5. Stiglitz, 'Globalization and Development', p. 63.
6. David Humphreys, *Logjam: Deforestation and the Crisis of Global Governance* (London: Earthscan, 2006), p. 37.
7. Lawrence Susskind, *Environmental Diplomacy: Negotiating More Effective Global Agreements* (New York, Oxford: Oxford University Press, 2004), p. 52.
8. Howlett and Rayner, 'Globalization and Governance Capacity', pp. 255–256.
9. Pieter Glasbergen, 'Learning to Manage the Environment', in *Democracy and the Environment: Problems and Prospects*, ed. William M. Lafferty and James Meadowcroft (Cheltenham and Lyme: Edward Elgar Publishing, 1996), pp. 175–190.
10. Skjærseth et al., 'Soft Law, Hard Law', p. 104.
11. Humphreys, *Forest Politics*, p. 63.
12. FSC, 'About the Forest Stewardship Council – FSC', http://www.fsc.org/about-fsc.html, accessed 03 November 2008.
13. PEFC, 'PEFC is the World's Largest Forest Certification Organisation', http://www.pefc.org/internet/html/, accessed 03 November 2008.
14. UN, 'United Nations Forum on Forests', http://www.un.org/esa/forests/, accessed 03 November 2008.
15. Canadian Standards Authority, 'About ISO/TC 207', http://www.TC207.org, accessed 10 May 2007.

16. Humphreys, *Logjam*, p. xvii.
17. Humphreys, *Forest Politics*, pp. 2–15.
18. Erik Hysing and Jan Olsson, 'Sustainability through good Advice? Assessing the Governance of Swedish Forest Biodiversity', *Environmental Politics* 14(4) (2005), pp. 510–526, at pp. 521–523; Howlett and Rayner, 'Globalization and Governance Capacity', pp. 251–252.

8 Conclusions

1. Humphreys, *Logjam*, pp. 11–14.
2. Peter Haas, 'UN Conferences and Constructivist Governance of the Environment', *Global Governance* 8(1) (2002), pp. 73–91, at p. 74, Glück et al., 'Changes in the Governance of Forest Resources', p. 51.
3. Robert Falkner, 'Private Environmental Governance and International Relations: Exploring the links', *Global Environmental Politics* 3(2) (2003), pp. 72–87, at p. 72.
4. Cashore et al., *Governing Through Markets*, pp. 27–29.
5. Lammerts van Beuren and Blom, *Hierarchical Framework*, p. 7. It is noted, however, that 'specific spatial levels may require additional principles particularly relevant to that level' (ibid.).
6. Lammerts van Beuren and Blom, *Hierarchical Framework*, p. 34.
7. 'A verifier is the source of information for the indicator, or for the reference value of the indicator' (Erik M. Lammerts van Beuren and Blom, *Hierarchical Framework*, p. 35.
8. Ibid., p. 25.
9. Ibid., p. 29.
10. I would like to thank Dr Fred Gale for raising the matter of intersubjectivity.
11. Thanks go once again to Dr Benjamin Cashore for this observation.
12. For more information, see: Charlie Parker, Andrew Mitchell, Mandar Trivedi and Niki Mardas, *The Little REDD-plus Book* (Oxford: Global Canopy Programme, 2009), pp. 11–92.
13. Ibid.; Allison Bleaney, Leo Peskett, and David Mwayafu, 'REDD-plus After Copenhagen: What Does it Mean on the Ground?' REDD-NET COP 15 Briefing, January 2010, http://6823165678770790248-a-redd—net-org-s-sites. googlegroups.com/a/redd-net.org/site/files/COP15outcome-webfinal.pdf, accessed 20 May 2010.
14. The Centre for People and Forests (RECOFT), *REDD-plus: Moving Forward for People and Forests* (Bangkok: RECOFT, 2010), http://www.recoftc.org/site/fileadmin/docs/publications/The_Grey_Zone/2010/REDD__2010_Moving FINAL.pdf, accessed 18 May 2010.
15. UN-REDD, 'The UN-REDD Programme' http://www.un-redd.org/UNREDD Programme/tabid/583/language/en-US/Default.aspx, accessed 23 March 2010, Forest Carbon Partnership Facility, 'Forest Carbon Partnership Facility', undated, pp. 6–10.
16. Andrew Jordan, Rüdiger K. W. Wurzel and Anthony Zito, 'The Rise of "New" Policy Instruments in Comparative Perspectives: Has Governance Eclipsed Government?' *Political Studies* 53 (2005), pp. 441–469, at pp. 492–494. Here the poles are governance–government, but the argument has

been extended to include state–non-state interests along a continuum of governance *authority*. The author would like to thank Dr Tek Maraseni of the Faculty of Business, University of Southern Queensland, for his assistance in preparing these data.

17. Interview with NGO participants at COP-15, 12 April 2010. The researcher was also present as an observer at the pre-Copenhagen climate change talks in Barcelona.

18. Volker Rittberger and Martin Nettesheim, 'Editors' Preface', in *Authority in the Global Economy*, ed. Volker Rittberger and Martin Nettesheim (Basingstoke: Palgrave Macmillan, 2008), pp. ix–xi, at pp. ix–x. 23 November 2010.

19. Volker Rittberger, Carmen Huckel, Lothar Rieth and Melanie Zimmer, 'Inclusive Global Institutions for a Global Political Economy', in *Authority in the Global Economy*, ed. Volker Rittberger and Martin Nettesheim (Basingstoke: Palgrave Macmillan, 2008), pp. 13–54, 14–15.

20. Ibid., p. 43.

21. Inge Kaul, 'Providing and Managing Public Goods', in *Authority in the Global Economy*, ed. Volker Rittberger and Martin Nettesheim (Basingstoke: Palgrave Macmillan, 2008), pp. 89–115, at p. 91.

22. Ibid., p. 102.

23. Ibid., pp. 104–105.

24. Peter-Tobias Stoll, 'Global Public Goods: The Governance Dimension', in *Authority in the Global Economy*, ed. Volker Rittberger and Martin Nettesheim (Basingstoke: Palgrave Macmillan, 2008), pp. 116–138, at p. 127.

25. Peter Utting and José Carlos Marques, 'Introduction: The Intellectual Crisis of CSR', in *Corporate Social Responsibility and Regulatory Governance: Towards Inclusive Government?*, ed. Peter Utting and José Carlos Marques (Basingstoke: Palgrave Macmillan, 2010), pp. 1–25, at pp. 10–11.

26. Doris Fuchs and Agni Kalfagiani, 'Private Food Governance: Implications for Social Sustainability and Democratic Legitimacy', in *Corporate Social Responsibility and Regulatory Governance*, ed. Peter Utting and José Carlos Marques (Basingstoke: Palgrave Macmillan, 2010), pp. 225–247, at pp. 240–241.

27. Florence Palpacuer, 'Challenging Global Governance in Global Commodity Chains', in *Corporate Social Responsibility and Regulatory Governance*, ed. Peter Utting and José Carlos Marques (Basingstoke: Palgrave Macmillan, 2010), pp. 276–299, at p. 295.

28. Peter Utting, 'Regulation in Global Governance', in *Authority in the Global Economy*, ed. Volker Rittberger and Martin Nettesheim (Basingstoke: Palgrave Macmillan, 2008), pp. 241–275, at p. 245.

29. Paddy Ireland and Renginee Pillay, 'Corporate Social responsibility in a Neoliberal Age', in *Corporate Social Responsibility and Regulatory Governance*, ed. Peter Utting and José Carlos Marques (Basingstoke: Palgrave Macmillan, 2010), pp. 77–104, at p. 95, citing various sources.

30. Jeffrey Atkinson and Martin Scurrah, *Globalizing Social Justice: The Role of Non-government Organizations in Bringing About Social Change 2009* (Basingstoke: Palgrave Macmillan, 2010), pp. 217–218.

31. Steven Bernstein and Benjamin Cashore, 'The Two-Level Logic of Non-state Market Driven Global Governance', in *Authority in the Global Economy*, ed. Volker Rittberger and Martin Nettesheim (Basingstoke: Palgrave Macmillan, 2008), pp. 276–313, at pp. 290–306.

32. Virginia Haufler, 'MNCs and the International Community: Conflict, Conflict Prevention and the Privatization of Diplomacy', in *Authority in the Global Economy*, ed. Volker Rittberger and Martin Nettesheim (Basingstoke: Palgrave Macmillan, 2008), pp. 217–240, at p. 232.
33. Bernstein and Cashore 'The Two-Level Logic', p. 306.

Bibliography

Accreditation Services International, 'Procedures for FSC Accreditation of Certification Bodies', ASI-PRO-20-110 Version 1.0, September 2006.

Acuña, Eliana, *The ECL Space Project. Learning from Social and Environmental Schemes for the ECL Space: ISO 14401 Case Study*. Pully: Pi Environmental Consulting, 2004.

Albertina, Skaidrite, 'Forest Owners Start to Sell Products from Certified Forests in Latvia', *PEFCC Newsletter* 10 (March 2002), p. 4.

Altoft, Katie, 'Canada – CSA', *PEFCC Newsletter* 12 (September, 2002), p. 8. ANEC/ECOS, 'Joint ANEC/ECOS Comments on the ISO 14000 Series Review', ANEC-ENV-G-030final (October, 2007).

ANEC/EEB, 'ANEC/EEB Position Paper on Environmental Management System Standards' (February, 2003).

Anonymous, *Certification in Indonesia*. London: Down to Earth and The Rainforest Foundation, 2001.

———, 'Pan European Forest Certification', memorandum, 2/9/199, http://www.faf.de/paneuro_e.htm, accessed 20 September 1999.

———, 'Stakeholder Involvement Efforts at the National Level: Summary and Analysis of the NSB Survey Results' (undated).

Arancibia, Daniel, *Forest Stewardship Standards in FSC System*. Pully: Pi Environmental Consulting and WWF, 2003.

Arnstein, Sherry, 'A Ladder of Citizen Participation', *Journal, American Institute of Planners* 35(4) (1969), pp. 216–224.

Articles and News, '3rd Meeting of ISO/TC 207 in Oslo, Norway', (August, 1995).

———, 'Communiqué from the 7th Annual TC 207 Plenary', ISO/TC 207 N357 1999-06-06 (July, 1999).

———, 'Environment – ISO/TC 207 Considers Industry's Needs', (August, 1995).

———, 'Summary Report – 2nd Meeting of ISO/TC 207 in Australia', (December, 1994).

Arts, Bas, 'Non-state Actors in Global Governance: New Arrangements Beyond the State', in *New Modes of Governance in the Global System: Exploring Publicness, Delegation and Inclusiveness*, eds, Mathias Koenig-Archibugi and Michael Zürn. Basingstoke: Palgrave Macmillan, 2006, pp. 177–200.

Atkinson, Jeffrey and Martin Scurrah, *Globalizing Social Justice: The Role of Non-government Organizations in Bringing About Social Change 2009*. Basingstoke: Palgrave Macmillan, 2010.

Azmat, Fara, 'Good Governance and Market-based Reforms: A Study of Bangladesh', *International Review of Administrative Sciences* 71(4) (2005), pp. 625–638.

Bäckstrand, Karin, 'Multi-Stakeholder Partnerships for Sustainable Development: Rethinking Legitimacy, Accountability and Effectiveness', *European Environment* 16 (2006), pp. 290–306.

Bäckstrand, Karin and Eva Lövbrand, 'Planting Trees to Mitigate Climate Change: Contested Discourses of Ecological Modernization, Green Governmentality and Civic Environmentalism', *Global Environmental Politics* 6(1) (2006), pp. 51–75.

Barrera, Gisella, 'New CERTFOR Standard for Native Forests', *PEFC News* 26 (July 2005).

Berger, Gerald, 'Reflections on Governance: Power Relations and Policy Making in Regional Sustainable Development', *Journal of Environmental Policy and Planning* 5(3) (2003), pp. 219–234.

Bernstein, Johanna, 'Sustainable Development Governance Challenges in the New Millennium', in *International Environmental Law-making and Diplomacy Review 2004*, ed., Marko Berglund. Joensuu: University of Joensuu Department of Law, 2005, pp. 31–50.

Bernstein, Steven, 'Liberal Environmentalism and Global Environmental Governance', *Global Environmental Politics* 2(3) (2002), pp. 1–16.

Bernstein, Steven and Benjamin Cashore, 'The Two-Level Logic of Non-state Market Driven Global Governance', in *Authority in the Global Economy*, eds, Volker Rittberger and Martin Nettesheim. Basingstoke: Palgrave Macmillan, 2008, pp. 276–313.

———, 'Nonstate Global Governance: Is Forest Certification a Legitimate Alternative to a Global Forest Convention?', in *Hard Choices, Soft Law: Combining Trade, Environment, and Social Cohesion in Global Governance*, eds, John J. Kirton and M. J. Trebilcock. Aldershot: Ashgate Press, 2004, pp. 33–64.

Bichsel, Anne, 'NGOs as Agents of Public Accountability and Democratisation in Intergovernmental Forums', in *Democracy and the Environment*, eds, William M. Lafferty and James Meadowcroft. Cheltenham and Lyme: Edward Elgar, 1996, pp. 234–255.

Birnie, Patricia, 'The UN and the Environment', in *United Nations, Divided World: The UN's Roles in International Relations*, eds, Adam Roberts and Benedict Kingsbury. Oxford: Oxford University Press, 2000, pp. 327–383.

Blomgren Bingham, Lisa, Rosemary O'Leary and Tina Nabatchi, 'Legal Frameworks for the New Governance: Processes for Citizen Participation in the Work of Government', *National Civic Review*, Spring (2005), pp. 54–61.

Boström, Magnus, 'How State-Dependent is a Non-State-Driven Rule-Making Project? The Case of Forest Certification in Sweden', *Journal of Environmental Policy & Planning* 5(2) (2003), pp. 165–180.

———, 'Regulatory Credibility and Authority through Inclusiveness: Standardization Organizations in Cases of Eco-labelling', *Organization* 13(3) (2006), pp. 345–367.

Boyer, Brook, 'Multilateral Negotiation Simulation Exercise: The Sustainable Management and Conservation of Forests', *International Environmental Law-making and Diplomacy Review 2005*, ed., Marko Berglund. Joensuu: University of Joensuu Press, 2005, pp. 299–310.

Bradbury, Roger, 'Are Indicators Yesterday's News?' in Tracking Progress: Linking Environment and Economy through Indicators and Accounting Systems. Conference Papers. 1996 Australian Academy of Science Fenner Conference on the Environment. Sydney: Institute of Environmental Studies, The University of New South Wales, 1996. Reproduced at http://www.tjurunga.com/biography/roger-papers.html, accessed 15 September 2007.

Briggs, Susan L. K., 'Do Environmental Management Systems Improve Performance?', *Quality Progress* 39(9) (2006), p. 78.

Cadman, Tim, 'Theory and Practice of Non-state Participation in Environmental and Forest-related Decision-making', in *International Environmental Law-making and Diplomacy Review 2005*, ed., Marko Berglund. Joensuu: University of Joensuu Department of Law, 2006, pp. 155–178.

Callanan, Mark, 'Institutionalizing Participation and Governance? New Participative Structures in Local Government in Ireland', *Public Administration*, 83(4) (2005), pp. 909–929.

Canadian Standards Authority, 'About ISO/TC 207', http://www.TC207.org, accessed 10 May 2007.

Capistrano, Doris, Markku Kanninen, Manuel Guariguata, Chris Barr, Terry Sunderland and David Raitzer, 'Revitalizing the UNFF: Critical Issues and Ways Forward', paper prepared for the Country-led Initiative on Multi-year Programme of Work of the United Nations Forum on Forests: Charting the Way Forward to 2015, Bali, Indonesia, 13–16 February 2007.

Caruso, Emily and Leontien Krul, 'Special FERN-FPP Report: UNFF Failing its Mandate – 4th Session of the United Nations Forum on Forests', *EU Forest Watch* June (2004), pp. 1–3.

Cashore, Benjamin, Fred Gale, Errol Meidinger and Deanna Newsom, *Confronting Sustainability: Forest Certification in Developing and Transitioning Countries*. New Haven: Yale Forestry and Environmental Studies Series, 2006.

Cashore, Benjamin, Graeme Auld and Deanna Newsom, *Governing Through Markets: Forest Certification and the Emergence of Non-State Authority*. New Haven and London: Yale University Press, 2004.

———, 'Forest Certification (Eco-labelling) Programs and their Policy-making Authority: Explaining Divergence Among North American and European Case Studies', *Forest Policy and Economics* 5 (2003), pp. 225–247.

Cashore, Benjamin, Elizabeth Egan Graeme Auld and Deanna Newsom, 'Revising Theories of Nonstate Market-Driven (NSMD) Governance: Lessons from the Finnish Forest Certification Experience', *Global Environmental Politics* 7(1) (2007), pp. 1–44.

Clapp, Jennifer, 'Global Environmental Governance for Corporate Responsibility and Accountability', *Global Environmental Politics* 5(3) (2005), pp. 23–34.

———, 'Standard Inequities', in *Voluntary Initiatives: The New Politics of Corporate Greening*, ed., Robert B. Gibson. Peterborough: Broadview Press, 1999.

———, 'The Privatization of Global Environmental Governance: ISO 14000 and the Developing World', *Global Governance* 4(3) (1998), pp. 295–316.

Coglianese, Cary, *Is Consensus an Appropriate Basis for Regulatory Policy?* Cambridge MA: Harvard University Press, 2000.

Combined New Zealand Environmental NGOs, 'Letter of Inquiry to Scientific Certification Systems (SCS) on Their Certification of Fletcher Challenge Forests (NZ)', dated April 2001.

Commonwealth of Australia, *Assessing the Sustainability of Forest Management in Australia*. Canberra: Department of Primary Industries and Energy, Forests Division and Montreal Implementation Group, undated.

———, *Australia's First Approximation Report For The Montreal Process*. Canberra: Montreal Implementation Group, 1997.

Counsell, Simon, *Trickery Or Truth? An Examination Of The Effectiveness Of The Forest Stewardship Council*. London: The Rainforest Foundation, 1999.

Counsell, Simon and Kim Terje Loraas, *Trading in Credibility: The Myth and Reality of the Forest Stewardship Council*. London and Oslø: Rainforest Foundation, 2002.

Courville, Sasha, 'Social Accountability Audits: Challenging or Defending Democratic Governance?', *Law and Policy* 25(3) (2003), pp. 269–297.

——, 'Understanding NGO-Based Social and Environmental Regulatory Systems: Why We Need New Models of Accountability', paper presented to the workshop Accountability in a Complex World: Conceptualizing Constitutional and Regulatory Efficacy in the Face of Globalization and Privatized Governance, 14–15 March 2003, New York, 23 pp.

Crowfoot, James E. and Wondolleck, Julia M., *Environmental Disputes: Community Involvement in Conflict Resolution*. Washington and Clovello: Island Press, 1990.

Department of Economic and Social Affairs/Secretariat of the United Nations Forum on Forests, Department of Economic and Social Affairs/UNFF Secretariat, 'Implementation of Proposals for Action Agreed by Intergovernmental Panel on Forests and by Intergovernmental Forum on Forests (IPF/IFF) Action for Sustainable Forest Management', December 2005.

——, 'Frequently Asked Questions', http://www.un.org/esa/forests/faq.html, accessed 15 March 2007.

——, 'United Nation Forum on Forests. Global Partnership: For Forests For People', Fact Sheet 1 (2004), http://www.un.org/esa/forests/factsheet.pdf, accessed 28 October 2004.

Dimitrov, Radoslav S., 'Hostage to Norms: States, Institutions and Global Forest Politics', *Global Environmental Politics* 5(4) (2005), pp. 1–24.

Dingwerth, Klaus, 'Private Transnational Governance and the Developing World: A Comparative Perspective', *International Studies Quarterly* 52 (2008), pp. 607–634.

Dobson, Andrew, 'Representative Democracy and the Environment', in *Democracy and the Environment*, eds, William M. Lafferty and James Meadowcroft. Cheltenham, and Lyme: Edward Elgar, 1996, pp. 124–139.

Donovan, Richard Z., Luis Fernando, Guedes Pinto and Lineu Siqueira, 'Withdrawal of Aracruz Riocell Guaíba Forest Management Certificate', letter to FSC stakeholders (24 May 2006).

Dorf, Michael and Charles Sabel, 'A Constitution of Democratic Experimentalism', *Columbia Law Review* 98(2) (1998), pp. 267–473.

Dryzek, John, *Discursive Democracy*. Cambridge: Cambridge University Press, 1990.

——, *The Politics of the Earth: Environmental Discourses*. New York: Oxford University Press, 1990.

Earth Negotiations Bulletin, 'Draft Decision for Adoption by ECOSOC', 13(133) (2005), p. 12.

——, 'A Brief Analysis of UNFF-5', 13(133) (2005), p. 14.

——, 'A Brief Analysis of UNFF-6', 13(144) (2006), p. 11.

——, 'Annex: MYPOW for 2007–2015', 13(162) (2007), p. 16.

——, 'Facing the Future', 13(174) (2007), pp. 9–10.

——, 'Ministerial Declaration', 13(133) (2005), p. 1.

——, 'Multi-stakeholder Dialogue', 13(133) (2005), p. 3.

——, 'Opening Statements', 13(144) (2006), p. 3.

——, 'Polite Listening: Engaging Civil Society', 13(116) (2004), pp. 9–10.

——, 'Quo Vadis: Contemplating the Post-UNFF Era', 13(116) (2004), p. 11.

——, 'Report of the Ad Hoc Expert Group on Parameters', 13(133) (2005), p. 4.

——, 'Summary of the Eighth Session of the United Nations Forum on Forests: 20 April–1 May 2009', 13(174) (2009), p. 1.

——, 'Summary of the Sixth Session of the United Nations Forum on Forests: 13–24 February 2006', 13(144) (2006), p. 1.

——, 'The Future of the International Arrangement on Forests', 13(144) (2006), p. 12.

——, 'The MPOW: Not Just Another Shopping List', 13(162) (2007), pp. 16–18.

——, 'The NLBI: Where's The Meat?', 13(162) (2007), p. 18.

——, 'The Role of UNFF and Other Players', 13(150) (2006), p. 11.

——, 'Things to Look For Today', 13(100) (2003), p. 2.

——, 'UNFF-5 Closing Statements', 13(133) (2005), p. 12.

——, 'Whither the Trees?' 13(144) (2006), p. 11.

Eberlein, Burkard and Dieter Kerwer, 'New Governance in the European Union: A Theoretical Perspective', *Journal of Common Market Studies* 42(1), pp. 121–142.

ECOSOC, 'Economic Commission for Europe, Committee for Trade, Industry and Enterprise Development, Working party on Technical Harmonization and Standardization Policies, Twelfth Session, 28–30 October 2002, Item 8(b) of the Provisional Agenda, Environmental Management Standards, Summary of the Main Results of the Tenth Annual Meeting of ISO/TC 207', Document TRADE/WP. 6/2002/12 (October, 2002).

——, 'Report of the Fourth Session of the Intergovernmental Forum on Forests', ECOSOC Resolution 2000/35 (18 October 2000), pp. 64–66.

——, '2006/49 Outcome of the Sixth Session of the United Nations Forum on Forests', ECOSOC Resolution 2006/49, E/2006/INF/2/Add.1, pp. 162–187.

Edwards, Brad, Jill Gravender, Annette Killmer, Genia Schenke and Mel Willis, *The Effectiveness of ISO 14001 in the United States*. Santa Barbara: Donald Bren School of Environmental Science & Management, University of California, 1999.

Edwards, Mark, 'Australia – A Change in the Woods', *PEFC News* 16 (October 2003), p. 3.

Elliott, Christopher, *Forest Certification: A Policy Perspective*. Bogor: Center for International Forestry Research, 2000.

EU Forest Watch, 'Presentation of the Pan European Forest Certification Scheme', 29 (December 1998) http://www.FERN.org, accessed 20 September 2002.

Evison, Irene, *FSC National Initiatives Manual*. Oaxaca: Forest Stewardship Council A.C., 1998.

Ezzy, Douglas, *Qualitative Analysis: Practice and Innovation*. Crows Nest: Allen & Unwin, 2002.

Falkner, Robert, 'Private Environmental Governance and International Relations: Exploring the Links', *Global Environmental Politics* 3(2) (2003) pp. 72–87.

Farrell, Katharine N., 'Recapturing Fugitive Power: Epistemology, Complexity and Democracy', *Local Environment* 9(5) (2003), pp. 469–470.

FERN, 'Major Group Participation in the UNFF What Does it Mean?' undated.

Finnish Forest Industries Federation, 'Finnish Forest Industries Need Raw Material from Certified Forests', media release 07 September 2000.

———, 'Increased credibility for Finnish Forest Certification Approach through European Co-operation', media release 25 August 1998.

———, 'Industry fully supports the Finnish Forest Certification System (FFCS) – FSC-Labelling of Products Becomes Available, Too?' media release 28 October 1998.

Fiorino, Daniel, 'Environmental Policy and the Participation Gap', in *Democracy and the Environment: Problems and Prospects*, eds, William M. Lafferty and James Meadowcroft. Cheltenham and Lyme: Edward Elgar, 1996, pp. 194–212.

———, 'Rethinking Environmental Regulation: Perspectives on Law and Governance', *The Harvard Environmental Law Review* 23(2) (1999), pp. 441–469.

Folke, Carl, Thomas Hahn, Per Olsson and Jon Norberg, 'Adaptive Governance of Social-Ecological Systems', *Annual Review of Environment and Resources*, 30 (2005), pp. 441–473.

Forest Carbon Partnership Facility, 'Forest Carbon Partnership Facility', undated.

Forest Stewardship Council, *10 years of FSC 1993–2003 Looking to the Future.* Bonn: Forest Stewardship Council A.C., 2004.

———, 'About FSC – FAQs', http://www.fsc.org/en/about/about_fsc/faqs, accessed 21 June 2007.

———, 'About FSC – History', http://www.fsc.org/en/about/about_fsc/history, accessed 21 June 2007.

———, 'About FSC – Services', http://www.fsc.org/en/about/about_fsc/services, accessed 21 June 2007.

———, 'About the Forest Stewardship Council – FSC', http://www.fsc.org/about-fsc.html, accessed 03 November 2008.

———, 'Accreditation', http://www.fsc.org/en/about/accreditation, accessed 21 June 2007.

———, 'Accreditation – Accreditation of Certification Bodies', http://www.fsc.org/en/about/accreditation/accred-certbod, accessed 21 June 2007.

———, 'Accreditation – Accreditation of Forest Stewardship Standards', http://www.fsc.org/en/about/accreditation/accred_fss, accessed 21 June 2007.

———, 'Accreditation – Accreditation of National Initiatives', http://www.fsc.org/en/about/accreditation/accred_ni, accessed 21 June 2007.

———, 'Announcement. Expanded Review and Revision of the FSC Principles and Criteria (FSC-STD-01-001 Version 4.0)' (undated).

———, *2001 Annual Report: Charting Our Future.* Oaxaca: Forest Stewardship Council A.C., undated.

———, *Annual Report 1999.* Oaxaca: Forest Stewardship Council A.C., 1999.

———, *Annual Report 2000.* Bonn: FSC International Center, undated.

———, 'Chain of Custody Certification Reports', Standard FSC-STD-20-010 Version 2.1 EN, November 2004.

———, 'Final Motions and Results from FSC General Assembly 2002' (undated).

———, 'Final Motions and Results From the FSC General Assembly 2008' (undated).

———, 'Forest Certification Reports', FSC Standard FSC-STD-20-008 Version 2.1, November 2004.

———, 'Forest Certification Public Summary Reports', FSC Standard FSC-STD-20-009, Version 2.1 EN, November 2004.

———, 'Forest Management Evaluation', FSC Standard FSC-STD-20-007 Version 2.1, November 2004.

———, 'Forest Pre-evaluation Visits', FSC Standard FSC-STD-20-005, Version 2.1, November 2004.

———, *Forest Stewardship Council 2000 Annual Report*. Oaxaca: Forest Stewardship Council, 2000.

———, 'Forest Stewardship Council A.C. By-Laws', Document 1.1 ratified September 1994; editorial revision, October 1996; revised February 1999, August 2000, November 2002, June 2005 and June 2006.

———, 'Forest Stewardship Council Interim Dispute Resolution Protocol', Document 1.4.3, adopted January 1998.

———, 'FSC Accredited Certification Bodies', Document 5.3.1, 6 June 2007.

———, 'FSC Accredited Forest Stewardship Standards', 30 March 2007.

———, 'FSC Chain of Custody Standard for Companies Supplying and Manufacturing FSC-Certified Products', FSC Standard FSC-STD-40-004, Version 1.0, September 2004.

———, 'FSC Controlled Wood Standard for Forest Management Enterprises', FSC Standard FSC-STD-30-010 Version 2.0 EN, October 2006.

———, 'FSC Governance Review Process Final Proposals Based on Consultative Process Held During 2007–2008', 3 October 2008.

———, 'FSC National Initiatives', Document 5.1.2, 5 June 2007.

———, 'FSC Plantations Review Policy Working Group Final Report', October 2006.

———, 'FSC Plantations Review Report from the First Policy Working Group meeting', March 2005.

———, 'FSC Principles and Criteria for Forest Stewardship' FSC Standard FSC-STD-01-001 Version 4.0, approved 1993, amended 1996, 1999 and 2002.

———, 'FSC Requirements for the Promotional Use of FSC Trademarks by FSC Certificate Holders and Non-certified Commercial Organizations', FSC Standard FSC-TMK-50-201, Version 1.0 EN, April 2007.

———, 'FSC Social Strategy: Building and Implementing a Social Agenda', version 2.1, February 2003.

———, 'FSC Standard for Forest Management Enterprises Supplying Non FSC-Certified Controlled Wood', FSC Standard FSC-STD-30-010, Version 1.0, September 2004.

———, 'FSC Standard for Non FSC-Certified Controlled Wood', FSC Standard FSC-STD-40-005, Version 1.0, September 2004.

———, 'General Requirements for FSC Accredited Certification Bodies: Application of ISO/IEC Guide 65:1996 (E)', FSC Standard FSC-STD-20-001, Version 2.1, November 2004.

———, 'Governance – Membership Chambers', http://www.fsc.org/en/about/governance/membership_chambers, accessed 21 June 2007.

———, 'Governance – Executive Director', http://www.fsc.org/en/about/governance/executive_director, accessed 21 June 2007.

———, 'FSC Governance Review', http://www.fsc.org/governance-review.html, accessed 15 January 2008.

———, 'Local Adaptation of Certification Body Generic Forest Stewardship Standards', FSC Standard FSC-STD-20-003 Version 2-1, November 2004.

———, 'Minutes of the Forest Stewardship Council General Assembly Oaxaca, Mexico, 24–25 June 1999'.

——, 'News', 8 June 2008, http://www.old.fsc.org/plantations.html, accessed 15 January 2009.

——, 'Qualifications for FSC Certification Body Auditors', FSC Standard FSC-STD-20-004 Version 2.2, November 2005.

——, 'Process for Developing FSC Forest Stewardship Standards', FSC Standard FSC-STD-60-006 Version 2.1, revised May 2005 and October 2005.

——, 'Response to the Rainforest Foundation Report "Trading in Credibility"', public statement dated 27 February 2003.

——, 'Review Process' http://www.old.fsc.org/plantations, accessed 10 July 2007.

——, 'Stakeholder Consultation for Forest Evaluation', FSC Standard FSC-STD-20-006, Version 2.1, November 2004.

——, 'Standard for Company Evaluation of FSC Controlled Wood', FSC Standard FSC-STD-40-005 Version 2.1 EN, October 2006.

——, 'Standard for Evaluation of FSC Controlled Wood in Forest Management Enterprises', FSC Standard FSC-STD-20-012 Version 1.0 EN, March 2007.

——, 'Statutes', Document 1.3, revised August 2000, November 2002 and June 2005.

——, 'Structure and Content of Forest Stewardship Standards', FSC Standard FSC-STD-20-002 Version 2.1 revised May 2005 and October 2005.

——, 'The Development and Approval of FSC International Standards', FSC Standard FSC-PRO-01-001 Version 1.0, November 2004.

FSC News and Notes, 'A First Step Forward for the Plantations Review', 2(8/9) (2004), p. 3.

FSC News and Notes Annual Review 2003, 3(1) (2004).

FSC News and Notes Annual Review 2006, 5(1) (2007).

FSC News and Notes, 'FSC Plantations Review Policy Working Group Completes its Task', 4(9) (2006), p. 1.

FSC News and Notes Annual Review 2005, 'Plantations, the Challenge Ahead', 4(1) (2006), p. 11.

Fuchs, Doris and Agni Kalfagiani, 'Private Food Governance: Implications for Social Sustainability and Democratic Legitimacy', in *Corporate Social Responsibility and Regulatory Governance: Towards Inclusive Government?* eds, Peter Utting and José Carlos Marques. Basingstoke: Palgrave Macmillan, 2010, pp. 225–247.

Gale, Fred, *The Tropical Timber Trade Regime*. Basingstoke: Macmillan Press, 1998.

Gale, Fred and Cheri Burda, 'The Pitfalls and Potential of Eco-certification as a Market Incentive for Sustainable Forest Management', in *The Wealth of Forests: Markets, Regulation, and Sustainable Forestry*, ed., Chris Tollefson. Paperback reprint. Vancouver: UBC Press, 1999, pp. 278–296.

Gaventa, John, 'Making Rights Real: Exploring Citizenship, Participation and Accountability', *Journal of International Development Studies* 33(2) (2002), pp. 1–11.

Geddes, Barbara, 'How the Cases You Choose Affect the Answers You Get: Selection Bias in Comparative Politics', *Political Analysis* 2 (1990), pp. 131–150.

Gill, Elisabeth, 'Norway – ENGO Withdraws from Revision Process', *PEFC News* 16 (October 2003), p. 7.

——, 'Norway – Living Forests Revision Process Started', *PEFCC Newsletter* 15 (May 2003), p. 5.

———, 'Unfounded Criticism of Norway's Forest Owners Federation by WWF', *PEFCC Newsletter* 8 (October 2001), pp. 3–4.

Glasbergen, Pieter, 'Learning to Manage the Environment', in *Democracy and the Environment: Problems and Prospects*, eds, William M. Lafferty and James Meadowcroft. Cheltenham and Lyme: Edward Elgar, 1996, pp. 175–193.

Glück, Peter, Jeremy Rayner, Benjamin Cashore, Arun Agrawal, Steven Bernstein, Doris Capistrano, Karl Hogl, Bernd-Markus Liss, Connie McDermott, Jagmohan S. Maini, Tapani Oksanen, Pekka Ollonqvist, Helga Pülzl, Rwald Rametsteiner and Werner Pleschberger, 'Changes in the Governance of Forest Resources', in *Forests in the Global Balance*, eds, G. Mery, R. Alfaro, M. Kaninnen, and M. Lobovikov. Helsinki: IUFRO, 2005, pp. 51–74.

Gulbrandsen, Lars H., 'Explaining Different Approaches to Voluntary Standards: A study of Forest Certification Choices in Norway and Sweden', *Journal of Environmental Policy and Planning* 7(1) (2005), pp. 43–59.

———, 'Overlapping Public and Private Governance: Can Forest Certification Fill the Gaps in the Global Forest Regime?' *Global Environmental Politics* 4(2) (2004), pp. 75–99.

———, 'Sustainable Forestry in Sweden: The Effect of Competition Among Private Certification Schemes', *The Journal of Environment & Development* 14(3) (2005), pp. 338–355.

———, 'The Effectiveness of Non-State Governance Schemes: A Comparative Study of Forest Certification in Norway and Sweden', *International Environmental Agreements* 5(2) (2005), pp. 125–149.

Gullison, R. E., 'Does Forest Certification Conserve Biodiversity?' *Oryx* 37(2) (2003), pp. 153–165.

Gunnerberg, Ben, 'Editorial', *PEFC News* 24 (February 2005), p. 1.

———, 'General Assembly Synopsis', *PEFCC News Special* (December 2002), p. 1.

———, 'PEFC Forging Ahead!', *PEFCC Newsletter* 11 (June 2002), p. 1.

Haas, Peter, M., 'Addressing the Global Governance Deficit', *Global Environmental Politics* 4(4) (2004), pp. 1–15.

———, 'International Institutions and Social Learning in the Management of Global Environmental Risks', *Policy Studies Journal* 28(3) (2000), pp. 558–575.

———, 'UN Conferences and Constructivist Governance of the Environment', *Global Governance* 8(1) (2002), pp. 73–91.

Habermas, Jürgen, *A Critique of Functionalist Reason (2)*. Cambridge: Polity Press, 1987.

———, *Between Facts and Norms: Contributions to a Discourse Theory of Law and Democracy*. Oxford: Blackwell, 1996.

Hague, Rod and Martin Harrop, *Comparative Government and Politics An Introduction*, 7th ed. Basingstoke: Palgrave Macmillan, 2007.

Hardin, Garrett, 'The Tragedy of the Commons', *Science*, 162(3859) (1968), pp. 1243–1248.

Haufler, Virginia, *A Public Role for the Private Sector: Industry Self-regulation*. Washington: Carnegie Endowment for International Peace, 2001.

———, 'MNCs and the International Community: Conflict, Conflict Prevention and the Privatization of Diplomacy', in *Authority in the Global Economy*, eds, Volker Rittberger and Martin Nettesheim. Basingstoke: Palgrave Macmillan, 2008, pp. 217–240.

Hauselmann, Pierre, *ISO Inside Out: ISO and Environmental Management*, 2nd ed. Godalming: WWF International, 1997.

Heaton, Kate, 'Smart Wood Program', in *International Conference on Certification and Labelling of Products from Sustainably Managed Forests*, ed., Lorraine Cairnes. Canberra: Australian Government Publishing Service, 1996, p. 69.

Held, David, 'Executive to Cosmopolitan Multilateralism', in *Taming Globalization: Frontiers of Governance*, eds, David Held and Mathias Koenig-Archibugi. Cambridge: Polity Press, 2003, pp. 174–177.

Held, David, Anthony G. McGrew, David Goldblatt and Jonathan Perraton, *Global Transformations: Politics, Economics and Culture*. London: Polity Press, 1999.

Herzog, Gabriele, 'Austria – More Than Three Quarters of Austria's Forests Certified', *PEFCC Newsletter* 9 (December 2001), pp. 6–7.

———, 'PEFC Austria – Totally Compliant with Austrian Law – Issues First Logo Licenses', *PEFCC Newsletter* 8 (July 2001), p. 5.

Hickey, Gordon M. and John L. Innes, 'Monitoring Sustainable Forest Management in Different Jurisdictions', *Environmental Monitoring and Assessment* 108 (2005), pp. 241–260.

Hirst, Paul, 'Democracy and Governance', in *Debating Governance: Authority, Steering and Democracy*, ed., Jon Pierre. Oxford and New York: Oxford University Press, 2000, pp. 13–35.

Hortensius, Dick, 'ISO 14000 and Forestry Management: ISO Develops "Bridging" Document', *ISO 9000-ISO 14000 NEWS* 4 (1999), pp. 11–20.

Howlett, Michael and Jeremy Rayner, 'Globalization and Governance Capacity: Explaining Divergence in National Forest Programs as Instances of "Next Generation" Regulation in Canada and Europe', *Governance*, 19(2) (2006), pp. 251–275.

Humphreys, David, *Forest Politics: The Evolution of International Cooperation*. London: Earthscan, 1996.

———, *Logjam: Deforestation and the Crisis of Global Governance*. London: Earthscan, 2006.

———, 'Redefining the Issues: NGO Influence on International Forest Negotiations', *Global Environmental Politics* 4(2) (2004), pp. 51–74.

———, 'The Certification Wars: Forest Certification Schemes as Sites for Trade-environment Conflicts', paper presented to the privatizing environmental governance panel at the 46th annual convention of the International Studies Association Honolulu, Hawaii, 1–5 March 2005.

Hysing, E. and Jan Olsson, 'Sustainability Through good Advice? Assessing the Governance of Swedish Forest Biodiversity', *Environmental Politics* 14(4) (2005), pp. 510–526.

IISD-ISO, 'A Background Paper to the International Organization for Standardization's (ISO) Strategic Advisory Group on Corporate Social Responsibility', CSR Briefing #1 (January 2003).

INNI, 'ISO 14000 and EMS INNI Articles', http://inni.pacinst.org/inni/EMS.htm#NGOCAG, accessed 28 October 2008.

———, 'ISO/TC 207 Approves Workplan to Improve NGO Participation', media release (18 January 2005), http://inni.pacinst.org/inni/NGO.htm, accessed 02 May 2007.

——, 'Drafting of ISO Social responsibility Standard Begins', media release (7 March 2006), http://inni.pacinst.org/inni/CSR.htm#SocialResponsibility, accessed 02 May 2007.

——, 'First Drafts of ISO's Water Management Standards Approved', media release (21 July 2005), http://inni.pacinst.org/inni/Water.htm#FinalizationOf, accessed 03 May 2007.

——, 'Improving the ISO 26000 Standard Development Process', media release (31 July 2006), http://inni.pacinst.org/inni/CSR.htm#SocialResponsibility, accessed 02 May 2007.

——, 'NGO Task Force Submits to TC207 Its First Deliverable to Improve Balanced Participation', media release 20 May 2007, http://inni.pacinst.org/inni/EMS.htm#NGOCAG, accessed 28 October 2008.

——, 'NGO Task Group Within ISO's Environmental Committee Develops Operational Guidance to Improve Stakeholder Involvement' media release (3 July 2006), http://inni.pacinst.org/inni/EMS.htm, accessed 02 May 2007.

——, 'Report from the ISO/TC 2007 2005 Plenary Meeting', media release (17 October 2005), http://inni.pacinst.org/inni/EMS.htm, accessed 02 May 2007.

Inoguchi, Takashi and Paul Bacon, 'Governance, Democracy, Consolidation and the "End of Transition"', *Japanese Journal of Political Science* 4(2) (2003), pp. 169–190.

Ireland, Paddy and Renginee Pillay, 'Corporate Social responsibility in a Neoliberal Age', in *Corporate Social Responsibility and Regulatory Governance: Towards Inclusive Government?*, eds, Peter Utting and José Carlos Marques. Basingstoke: Palgrave Macmillan, 2010, pp. 77–104.

ISEAL, 'Setting Social and Environmental Standards: A Research Report on Existing Standard-setting Practices, R028 – Public Draft 1, August 2003', http://www.isealalliance.org, accessed 26 October 2004.

ISO, 'Committee Structure', http://www.TC207.org/struc_detail.asp?group=SC7, accessed 18 June 2008.

——, 'How ISO Decides What Standards are Developed', http://www.iso.org/iso/en/aboutiso/introduction/index.html, accessed 02 May 2007.

——, 'How the ISO system is financed', http://www.iso.org/iso/en/aboutiso/introduction/index.html, accessed 02 May 2007.

——, 'How the ISO system is managed', http://www.iso.org/iso/en/aboutiso/introduction/index.html, accessed 02 May 2007.

——, *ISO 14001 Environmental Management Systems – Specifications with Guidance for Use*. Geneva: ISO, 1996.

——, 'ISO's membership rises to 150 countries', media release (20 April 2005), http://www.iso.org/iso/pressrelease.htm?refid=ref955, accessed 19 June 2008.

——, 'ISO's Structure', http://www.iso.org/iso/structure, accessed 18 June 2008.

——, *My ISO Job: Guidance For Delegates and Experts*. Geneva: ISO Central Secretariat, 2005.

——, 'Overview of the ISO system', http://www.iso.org/iso/en/aboutiso/introduction/index.html, accessed 24 April 2007.

——, 'Stages in the Development of International Standards', http://www.TC207.org, accessed 10 May 2007.

——, *Statutes and Rules of Procedure*, 14th ed. Geneva: ISO, 2000.

——, 'Who can join ISO', http://www.iso.org/iso/en/aboutiso/introduction/index.html, accessed 02 May 2007.

ISO/TC 207, 'Business Plan ISO/TC 207 Environmental Management', N726R0/CAG N 376R3 Version 3 (June, 2005).

———, 'Communiqué 6th Annual Meeting of ISO/TC 207 on Environmental Management', N283 (June, 1998).

———, 'Communiqué 8th Annual Meeting of ISO/TC 207 on Environmental Management', N429 REV 01:2000-07-10 (June, 2000).

———, 'Communiqué 13th Annual Meeting of ISO/TC 207 on Environmental Management', CAG N386 (September, 2005).

———, 'Draft ISO/TC 207 Proposed Process to Identify Stakeholder Balance at ISO/TC 207 Meetings', ISO/TC 207 CAG N439R1 (undated).

———, 'Draft: Recommendations for an Improved Balance of Stakeholder Participation in ISO TC 207', CAG N437R1 (April, 2007).

———, 'Increasing the Effectiveness of NGO Participation in ISO TC 207', N590 Rev/ISO/TC NGO TG N28 (June, 2003).

———, 'ISO's Membership Rises to 150 Countries', media release (20 April 2005), http://www.iso.org/iso/pressrelease.htm?refid=ref955, accessed 19 June 2008.

ISO/TC 207 NGO-CAG Task Force, 'Phase 1 of Workplan', Interim Report Final Draft (September, 2004).

Jänicke, Martin, 'Conditions for Environmental Policy Success: An International Comparison', *The Environmentalist* 12 (1992), pp. 47–58.

———, 'Democracy as a Condition for Environmental Policy Success: The Importance of Non-institutional Factors', in *Democracy and the Environment: Problems and Prospects*, eds, William M. Lafferty and James Meadowcroft. Cheltenham and Lyme: Edward Elgar, 1996, pp. 71–85.

Jordan, Andrew, Rüdiger K. W. Wurzel and Anthony Zito, 'The Rise of "New" Policy Instruments in Comparative Perspectives: Has Governance Eclipsed Government?' *Political Studies* 53 (2005), pp. 441–469.

Kaivola, Auvo, 'Finland – First Revised SFM Version Expected in November', *PEFCC Newsletter* 12 (September 2002), p. 6.

———, 'Ongoing Debate about Forestry and Reindeer Husbandry in Upper Lapland in Finland', *PEFC News* 25 (April 2005), p. 8.

———, 'Revision of Forest Certification Requirements Under Way in Finland', *PEFCC Newsletter* 11 (June 2002), p. 7.

Kaul, Inge, 'Providing and Managing Public Goods', in *Authority in the Global Economy*, eds, Volker Rittberger and Martin Nettesheim. Basingstoke: Palgrave Macmillan, 2008, pp. 89–115.

Keck, Margaret E. and Kathryn Sikkink, *Activists Beyond Borders: Advocacy Networks in International Politics*. New York: Cornell University Press, 1998.

Keohane, Robert O., 'Global Governance and Democratic Accountability', in *Taming Globalization: Frontiers of Governance*, eds, David held and Mathias Koenig-Archibugi. Cambridge: Polity Press, 2003, pp. 130–159.

Kerwer, Dieter, 'Governing Financial Markets by International Standards', in *New Modes of Governance in the Global System: Exploring Publicness, Delegation and Inclusiveness*, eds, Mathias Koenig-Archibugi and Michael Zürn. Basingstoke: Palgrave Macmillan, 2006.

King, Andrew A., Michael J. Lenox, Ann Terlaak, 'The Strategic Use of Decentralized Institutions: Exploring Certification with the ISO 14001 Management Standard', *Academy of Management Journal* 48(6) (2005), pp. 1091–1106.

Kjaer, Anne Mette, *Governance*. Cambridge and Malden MA: Polity Press, 2004.

Koenig-Archibugi, Mathias, 'Introduction: Institutional Diversity in Global Governance', in *New Modes of Governance in the Global System: Exploring Publicness, Delegation and Inclusiveness*, eds, Mathias Koenig-Archibugi and Michael Zürn. Basingstoke: Palgrave Macmillan, 2006, pp. 1–30.

Kölliker, A., 'Conclusion 1: Governance and Public Goods Theory', in *New Modes of Governance in the Global System: Exploring Publicness, Delegation and Inclusiveness*, eds, Mathias Koenig-Archibugi and Michael Zürn. Basingstoke: Palgrave Macmillan, 2006, pp. 201–235.

Kollman Kelly and Aseem Prakash, 'EMS-based Environmental Regimes as Club Goods: Examining Variations in Firm-level Adoption of ISO 14001 and EMAS in U.K., U.S. and Germany', *Policy Sciences* 35 (2002), pp. 43–67.

Kooiman, Jan, 'Findings, Speculations and Recommendations in *Modern Governance: New Government Society Interactions*, ed., Jan Kooiman. London: Sage, 1993, pp. 249–262.

———, 'Social-Political Governance: Introduction', in *Modern Governance: New Government Society Interactions*, ed., Jan Kooiman. London: Sage, 1993, pp. 1–8.

———, 'Societal Governance: Levels, Models, and Orders of Social-Political Interaction', in *Debating Governance: Authority, Steering and Democracy*, ed., Jon Pierre. Oxford: Oxford University Press, 2000, pp. 138–166.

Korsbakken, Ivar and Svein Søgnen, 'Norwegian Forest Owners Warn of Trade Barriers', *PEFC News* 35 (January 2007), p. 11.

Krul, Leontien, Tom Griffiths and Saskia Ozinga, 'Live or Let Die? An Evaluation of the Fifth Session of the United Nations Forum on Forests', *FERN Special Report*, July 2005, p. 3.

Lafferty, William M. and James Meadowcroft, 'Democracy and the Environment: Prospects for Greater Congruence', in *Democracy and the Environment: Problems and Prospects*, eds, William M. Lafferty and James Meadowcroft. Cheltenham and Lyme: Edward Elgar, 1996, pp. 256–272.

Lammerts van Beuren Erik M. and Esther M. Blom, *Hierarchical Framework for the Formulation of Sustainable Forest Management Standards*. Leiden: The Tropenbos Foundation, 1997.

Leslie, Andrew D., 'The Impacts and Mechanics of Certification', *International Forestry Review* 6(1) (2004), pp. 30–39.

Liimatainen, Matti and Sini Harkki, *Anything Goes? Report on PEFC Certified Forestry*. Helsinki: Greenpeace Nordic and the Finnish Nature League, 2001.

Lindström, Tommy, Eric Hansen and Heikki Juslin, 'Forest Certification: The View from Europe's NIPFs', *Journal of Forestry* 97(3) (1999), pp. 25–30.

Link, Sharon and Eitan Naveh, 'Standardization and Discretion: Does the Environmental Standard ISO 14001 Lead to Performance Benefits?' *IEEE Transactions on Engineering Management* 53(4) (2006), pp. 508–519.

Lipschutz, Ronnie D. and Judith Mayer, *Global Civil Society: The Politics of Nature from Place to Planet*. New York: State University of New York Press, 1996.

Mackendrick, Norah A., 'The Role of the State in Voluntary Environmental Reform: A Case Study of Public Land', *Policy Sciences* 38 (2005), pp. 21–44.

Mankin, Bill, 'MY POW or Yours? Choosing a Leadership Agenda for the UNFF', unpublished paper prepared for CIFOR, January 2007.

Mäntyranta, Hannes, *Forest Certification – An Ideal That Became an Absolute*, trans. Heli Mäntyranta. Helsinki: Metsälehti Kustannus, 2002.

Mason, Michael, *Environmental Democracy*. New York: St Martin's Press, 1999.

Mastenbroek, Ellen, 'EU Compliance: Still a "Black Hole"?', *Journal of European Public Policy* 12(6) (2005), pp. 1103–1120.

May, Peter, 'Regulation and Compliance Motivations: Examining Different Approaches', *Public Administration Review* 65(1) (2005), pp. 31–44.

MCPFE, 'Report on the Follow-Up of the Strasbourg and Helsinki Ministerial Conferences on the Protection of Forests in Europe', *Follow-up Reports on the Ministerial Conferences on the Protection of Forests in Europe*, ed., Liaison Unit in Lisbon, Vol. 1. Lisbon: Ministry of Agriculture, Rural Development and Fisheries of Portugal, 1998a.

———, 'Sustainable Forest Management in Europe, Special Report on the Follow-up on the Implementation of Resolutions H1 and H2 of the Helsinki Ministerial Conference', *Follow-up Reports on the Ministerial Conferences on the Protection of Forests in Europe*, ed., Liaison Unit in Lisbon, Vol. 2. Lisbon: Ministry of Agriculture, Rural Development and Fisheries of Portugal, 1998b.

Meidinger, Errol, 'Forest Certification as a Global Civil Society Regulatory Institution', in *Social and Political Dimensions of Forest Certification*, ed., Errol Meidinger, Christopher Elliott, and Gerhard Oesten. Remagen-Oberwinter: Forstbuch, 2003, pp. 265–292.

———, 'The Administrative Law of Global Private-Public Regulation: The Case of Forestry', *European Journal of International Law* 17(1) (2006), pp. 47–87.

Montréal Process, *Criteria and Indicators for the Conservation and Sustainable Management of Temperate and Boreal Forest Ecosystems*, 2nd ed. Unknown location: Montréal Process, 1999.

Morikawa, Mari and Jason Morrison, *Who develops ISO Standards? A Survey of Participation in ISO's International Standards Development Processes*. Oakland: Pacific Institute, 2004.

Nanz, Patrizia and Steffek, Jens, 'Assessing the Democratic Quality of Deliberation in International Governance: Criteria and Research Strategies', *Acta Politica*, 40(3) (2005), pp. 368–383.

Newman, Janet, Marian Barnes, Helen Sullivan and Andrew Knops, 'Public Participation and Collaborative Governance', *Journal of Social Politics* 33(2) (2004), pp. 203–223.

Nussbaum, Ruth and Markku Simula, *Forest Certification Handbook*. London: Earthscan, 2005.

Oberthür, Sebastian, Matthias Buck, Sebastian Müller, Stefanie Pfahl, Richard G. Tarasofsky, Jacob Werksman and Alice Palmer, *Participation of Non-governmental Organisations in International Environmental Governance: Legal Basis and Practical Experience*. Berlin: Ecologic/Centre for International and European Environmental Research, 2002.

O'Donnell, Guillermo, *Delegative Democracy. Counterpoints: Selected Essays on Authoritarianism and Democratization*. Notre Dame: University of Notre Dame Press, 1999, pp. 159–174.

O'Dwyer, Conor and Daniel Ziblatt, 'Does Decentralisation Make Government More Efficient and Effective?' *Commonwealth & Comparative Politics* 44(3) (2006), pp. 326–343.

Olsen, Tanja, 'PEFC Certification Denmark', *PEFC News* 23 (December 2004), p. 5.

Ostrom, Elinor, *Governing the Commons: The Evolution of Institutions for Collective Action*. Cambridge: Cambridge University Press, 1990.

Overdevest, Christine, 'Codes of Conduct and Standard Setting in the Forest Sector Constructing Markets for Democracy?' *Relations Industrielles/Industrial Relations* 59(1) (2004), pp. 172–197.

Ozinga, Saskia, *Behind the Logo: An Environmental and Social Assessment of Forest Certification Schemes*. Moreton-in-Marsh: Fern, 2001.

——, *Footprints in the Forest: Current Practice and Future Challenges in Forest Certification*. Moreton-in-Marsh: FERN, 2004.

——, 'The European NGO Position on the PEFC', presentation on behalf of FERN to the Pan-European Forest Certification (PEFC) Seminar and Workshop, Würzburg, 19–21 April 1999.

Palmujoki, Eero, 'Public-private Governance Patterns and Environmental Sustainability', *Environment, Development and Sustainability* 8 (2006), pp. 1–17.

Palpacuer, Florence, 'Challenging Global Governance in Global Commodity Chains', in *Corporate Social Responsibility and Regulatory Governance: Towards Inclusive Government?* eds, Peter Utting and José Carlos Marques. Basingstoke: Palgrave Macmillan, 2010, pp. 276–299.

Parker, Charlie, Andrew Mitchell, Mandar Trivedi and Niki Mardas, *The Little REDD + Book*. Oxford: Global Canopy Programme, 2009.

Parto, Saeed, 'Aiming Low', in *Voluntary Initiatives: The New Politics of Corporate Greening*, ed., Robert B. Gibson. Peterboroush: Broadview Press, 1999.

Pattberg, Phillip H., 'The Forest Stewardship Council: Risk and Potential of Private Forest Governance', *Journal of Environment and Development* 14(3) (2005), pp. 356–374.

——, 'The Institutionalization of Private Governance: How Business and Non-profit Organizations Agree on Transnational Rules', *Governance: An international Journal of Policy, Administration, and Institutions* 18(4) (2005), pp. 589–610.

——, 'What Role for Private Rule-Making in Global Environmental Governance? Analysing the Forest Stewardship Council (FSC)', *International Environmental Agreements* 5(2) (2005), pp. 175–189.

PEFC, *Annual Review 2006*. Luxembourg: PEFC, undated.

——, 'A Short History', http://www.pefc.org/internet/html/about_PEFC/4_1137_498.htm accessed 14 June 2007.

——, 'Basis for Certification Schemes and their Implementation', Annex 3, 29 October 2004.

——, 'Certification and Accreditation Procedures', Annex 6, 11 April 2005.

——, 'Chain of Custody of Forest Based Products – Requirements', Annex 4, 17 June 2005.

——, 'Endorsement and Mutual Recognition of National Schemes and their Revision', Annex 7, 29 October 2004.

——, 'PEFC Council Minimum Requirements Checklist', GL 2/2005, revised 28 April 2005.

——, 'PEFC Council Position Statement on Recent WWF Press Release on Austrian Forest Certification Scheme', media release 08 May 2001.

——, 'PEFC Council Procedures for the Investigation and Resolution of Complaints and Appeals', 28 June 2007.

——, 'PEFC Council Statutes', As Approved at General Assembly 22 November 2002.

——, 'PEFC Council Technical Document', 29 October 2004.

——, 'PEFC in Figures', http://www.pefc.org/internet/html/about_pefc/4_1137_515.htm accessed 14 June 2007.

———, 'PEFC is the World's Largest Forest Certification Organisation', http://www.pefc.org/internet/html/accessed 03 November 2008.

———, 'Rules for Standards Setting', Annex 2, 29 October 2004.

PEFCC Newsletter, 'Globalisation', 14 (February 2003), p. 1.

———, 'Guidelines Revision', 14 (February 2003), p. 2.

———, 'Revised PEFCC Guidelines Approved', *PEFCC Newsletter* 15 (May 2003), p. 4.

———, 'PEFC Austria is on the Road to Success – WWF Accusations "Substantially Wrong"', 7 (July 2001), p. 5.

———, 'PEFC Chairman to Visit Gabon to Discuss Feasibility of Pan African Forest Certification System', (December 2002), p. 3.

———, 'Some New Faces on the PEFC Council', 7 (July 2001), p. 2.

———, 'Statistics on Main Forest Certification Schemes', 7 (July 2001), p. 1.

PEFCC News Special, 'Continuing Globalisation of the PEFC Council', (December 2002), p. 2.

———, 'Did you Know? PEFC Wants People to Manage Forests Sustainably', fact sheet (April 2007), p. 3.

———, 'Position Paper: Tribal and Indigenous People, Local People, Local Communities, Forest Dependent Communities and the PEFC Council', November 2005.

PEFC Germany, 'Position of PEFC Germany Concerning the FERN-Report "Behind the Logo"', http://www.pefc.de/vergliech/behind.htm accessed 19 June 2007.

PEFC News, 'Alteration to the PEFC Name', 17 (November 2003), p. 4.

———, 'Assessment of Forest Certification Systems', 37 (May 2007), p. 3.

———, 'Australia – A Change in the Woods', 16 (October 2003), p. 3.

———, 'Belgian Public Procurement Policy Includes PEFC', 31 (May 2006), p. 2.

———, 'Election of New Board of Directors', 17 (November 2003), pp. 5–8.

———, 'ENGO Conference', 18 (February 2004), p. 3.

———, 'EU Parliament Regards PEFC and FSC as Equally Suitable', 30 (March 2005), p. 1.

———, 'European Toy Producers Treat PEFC and FSC Equally', 33 (October 2006), p. 1.

———, 'First ENGO Symposium', 20 (May 2004), p. 1.

———, 'German Government Specifies PEFC', 36 (March 2007), p. 1.

———, 'Governmental Shopper's Guides Recommend PEFC', 33 (October 2006), p. 2.

———, 'International Accreditation Forum', 19 (March 2004), p. 1.

———, 'International Chain of Custody', 18 (February 2004), p. 1.

———, 'International ENGO Platform Supporting PEFC Certification', 25 (April 2005), p. 2.

———, 'Japanese Public Procurement Policy Chooses PEFC', 31 (May 2006), p. 1.

———, 'Key Note Speech by Peter Seligmann from Conservation International', 34 (November 2006), p. 2.

———, 'New International PEFC Chain of Custody Released', 22 (November 2004), p. 7.

———, 'PEFC Chosen for French Government Timber procurement Policy', 25 (April 2005), p. 1.

———, 'PEFC Council Global Statistics', 37 (May 2007), p. 3.

————, 'PEFC Granted Observer Status in MCPFE', 35 (January 2007), p. 4.

————, 'PEFC: Only European? Based On Merely a Handful of Criteria? Not For Indigenous People? No Social Dialogue...?', 37 (May 2007), p. 2.

————, 'PEFC Open Letter To The World Bank And WWF', 21 (September 2004), p. 1.

————, 'PEFC Praised for its Social Criteria', 33 (October 2006), p. 1.

————, 'PEFC Sweden Publishes Policy on Sámi Reindeer Herding', 31 (May 2006), p. 1.

————, 'UK Government Confirms PEFC as Source of Legal and Sustainable Timber', 35 (January 2007), p. 1.

————, 'UK Government Timber Procurement Policy Continues to Recognize PEFC Certified Timber', 26 (July 2005), p. 3.

————, 'PEFC – World's Largest Resource of Certified Wood', 29 (January 2006), p. 1.

Perrons, Dianne, *Globalization and Social Change: People and Places in a Divided World*. London: Routledge, 2004.

Persson, R., 'Where is the United Nations Forum on Forests Going?' *International Forestry Review*, 7(4) (2005), pp. 348–357.

Peters, B. Guy, 'Governance and Comparative Politics', in *Debating Governance: Authority, Steering and Democracy*, ed., Jon Pierre. Oxford: Oxford University Press, 2000, pp. 37–51.

Pierre, Jon, and B. Guy Peters, *Governance, Politics and the State*. Basingstoke: Macmillan, 2000.

Plauche-Gillon, Henri, 'Extracts from the Chairman's Report on Activities in 2003/2004', *PEFC News* 22 (November 2004), p. 3.

Potoski, Matthew and Aseem Prakash, 'Green Clubs and Voluntary Governance: ISO 14001 and Firms' Regulatory Compliance', *American Journal of Political Science* 49(2) (2005), pp. 235–248.

————, 'Regulatory Convergence in Nongovernmental Regimes? Cross-National Adoption of ISO 14001 Certifications', *The Journal of Politics* 66(3) (2004), pp. 885–905.

Price, Trevor J., 'ISO 14001: Transition to Champion?' *Environmental Quality Management*, Spring (2007).

Raines, Susan Summers, 'Judicious Incentives: International Public Policy Responses to the Globalization of Environmental Management', *Review of Policy Research* 23(2) (2006), pp. 473–490.

————, 'Perceptions of Legitimacy and Efficacy in International Environmental Management Standards: The Impact of the Participation Gap', *Global Environmental Politics* 3(3) (2003), pp. 47–78.

Rehbinder, Eckhard, 'Forest Certification and Environmental Law', in *Social and Political Dimensions of Forest Certification*, eds, Errol Meidinger, Christopher Elliott, and Gerhard Oesten. Remagen-Oberwinter: Forstbuch, 2003, pp. 331–351.

Reuben, William, 'The Role of Civic Engagement and Social Accountability in the Governance Equation', *Social Development Notes* 75 (2003).

Rezende, Maria Teresa, 'Forum of Environmental NGOs and Brazilian Industries', *PEFC News* 36 (March 2007), p. 5.

Rhodes, Rod A. W., 'The New Governance: Governing Without Government', *Political Studies* 44 (1996), pp. 652–667.

———, *Understanding Governance: Policy Networks, Governance, Reflexivity and Accountability.* Buckingham: Open University Press, 1997.

Rittberger, Volker, Carmen Huckel, Lothar Rieth and Melanie Zimmer, 'Inclusive Global Institutions for a Global Political Economy', in *Authority in the Global Economy*, eds, Volker Rittberger and Martin Nettesheim. Basingstoke: Palgrave Macmillan, 2008, pp. 13–54.

Rittberger, Volker and Martin Nettesheim, 'Editors' Preface', in *Authority in the Global Economy*, eds, Volker Rittberger and Martin Nettesheim. Basingstoke: Palgrave Macmillan, 2008, pp. ix–xi.

Robinson, Dawn and Brown, Larianne, *The SLIMFs Initiative: A Progress Report. Increasing Access to FSC Certification for Small and Low Intensity Managed Forests.* Oaxaca: Forest Stewardship Council, 2002.

Romeijn, Paul, *Green Gold: On Variations in Truth in Plantation Forestry.* Heelsum: Treemail, 1999.

Rosenau, James, 'Change, Complexity and Governance in a Globalising Space', in *Debating Governance: Authority, Steering and Democracy*, eds, Jon Pierre. Oxford and New York: Oxford University Press, 2000, pp. 167–200.

———, *Distant Proximities: Dynamics Beyond Globalization.* Princeton and Oxford: Princeton University Press, 2003.

Royal Forest and Bird Protection Society, Greenpeace NZ, World Wide Fund For Nature NZ, Environment and Conservation Organisations of NZ, Native Forest Action, Friends of the Earth NZ, Native Forest Network – Southern Hemisphere, Federated Mountain Clubs of NZ 'Letter to FSC Accredited Forest Management Certifiers', dated 12 March 2001.

Ruggie, John G., 'Taking Embedded Liberalism Global: The Corporate Connection', in *Taming Globalisation: Frontiers of Governance*, eds, David Held and Mathias Koenig-Archibugi. Cambridge: Polity Press, 2003.

Salomon, Lester, 'The New Governance and the Tools of Public Action: An Introduction', *The Tools of Government: A Guide to the New Governance*, ed., Lester Saloman. Oxford: Oxford University Press, 2002.

Savcor Indufor Oy, *Effectiveness and Efficiency of FSC and PEFC Forest Certification on Pilot Areas in Nordic Countries.* Helsinki: Federation of Nordic Forest Owners' Organisations, 2005.

Scholte, Jan A., 'Civil Society and Democratically Accountable Global Governance', *Government and Opposition* 39 (2004), pp. 211–233.

Schutzgemeinschaft Deutscher Wald Landesverband Bayern e.V., *Jahresbericht 2005.*

Scrase, Hannah and Anders Lindhe, *Developing Forest Stewardship Standards – A Survival Guide.* Jokkmokk: Taiga Rescue Network, 2001.

Shannon, Margaret, 'What is Meant by Public Participation in Forest Certification Processes? Understanding Forest Certification Within Democratic Governance Institutions', in *Social and Political Dimensions of Forest Certification*, eds, Errol Meiginger, Christopher Elliott and Gerhard Oesten. Remagen-Oberwinter: Forstbuch, 2003, pp. 179–198.

Simmons, Richard and Johnston Birchall, 'A Joined-up Approach to user Participation in Public Services: Strengthening the "Participation Chain"', *Social Policy and Administration* 39(3) (2005), pp. 260–283.

Simula, Anna-Leena, 'Finnish Forest certification System (FFCS) and its Compatibility with International Certification Systems', presentation on behalf of MTK

Forestry Group to the Pan-European Forest Certification (PEFC) Seminar and Workshop, Würzburg, 19–21 April 1999.

Simula, Markku, 'Key Elements in the International certification Procedures: Review of the Pan European Certification', paper presented to the Pan-European Forest Certification (PEFC) Seminar and Workshop, Würzburg, 19–21 April 1999.

Skjærseth, Jon Birger, Olav Schram Stokke and Jørgen Wettestad, 'Soft Law, Hard Law, and Effective Implementation', *Global Environmental Politics* 6(3) (2006), pp. 104–120.

Slob, Bart and Gerard Oonk, *The ISO Working Group on Social Responsibility: Developing the Future ISO SR 26000 Standard*. Amsterdam: SOMO/Centre for Research on Multinational Corporations, 2007.

Smismans, Stijn, *Law, Legitimacy, and European Governance*. Oxford: Oxford University Press, 2004.

Sonnenfeld, David and Arthur Mol, 'Globalization and the Transformation of Environmental Governance: An Introduction', *American Behavioral Scientist*, 45(9) (2002), pp. 1318–1339.

Spickard, James, 'Human Rights, Religious Conflict, and Globalisation: Ultimate Values in a New World Order', *International Journal on Multicultural Societies* 1(1) (1999), pp. 2–20.

Stiglitz, Joseph E., 'Globalization and Development', in *Taming Globalisation: Frontiers of Governance*, eds, David Held and Mathias Koenig-Archibugi. Cambridge: Polity Press, 2003, pp. 47–67.

Stoker, Gerry, 'Governance as Theory: Five Propositions', *International Social Science Journal*, 50(155) (1998), pp. 17–28.

Stoll, Peter-Tobias, 'Global Public Goods: The Governance Dimension', in *Authority in the Global Economy*, eds, Volker Rittberger and Martin Nettesheim. Basingstoke: Palgrave Macmillan, 2008, pp. 116–138.

Susskind, Lawrence, *Environmental Diplomacy: Negotiating More Effective Global Agreements*. New York and Oxford: Oxford University Press, 2004.

Széll, Patrick, 'Introduction to the Discussion on Compliance', *International Environmental Law-making and Diplomacy Review 2004*, ed., Marko Berglund. Joensuu: University of Joensuu Department of Law, 2005, pp. 117–123.

Takeuchi, Haruyoshi, 'PEFC Asia Promotions Initiative', *PEFC News* 18 (February 2004), p. 3.

Tambini, Damian, 'Post-national Citizenship', *Ethnic and Racial Studies* 24(2) (2001), pp. 195–217.

Teegelbekkers, Dirk, 'Germany – Chain-of-Custody', *PEFC News* 16 (October 2003), p. 5.

———, 'Chain-of-Custody Certificates Increasing in Germany', *PEFCC Newsletter* 14 (February 2003), p. 6.

———, 'ENGOs Invited to Witness the Auditing Process in Germany', *PEFCC Newsletter* 7 (July 2001), pp. 5–6.

———, 'Forest Certification – Experiences With PEFC in Germany' presentation on behalf of PEFC Germany to the Seminar on Strategies for the Sound Use of Wood, Poiana Brasov, Romania, 24–27 March 2003.

———, 'Germany – Forest Owners Protest Against Discrimination', *PEFCC Newsletter* 11 (June 2002), p. 5.

————, 'Germany – Over 5 Million Hectares Already Certified', *PEFCC Newsletter*, 9 (December, 2001), p. 5.

The Wilderness Society, Certifying the Incredible, The Australian Forestry Standard. Barely Legal and not Sustainable. No location: The Wilderness Society, 2005.

Thinnes, Michèle, 'ENGOs Join PEFC', *PEFC News* 33 (October 2006), p. 4.

Thornber, Kirsti, 'Certification: A Discussion of Equity Issues', in *Social and Political Dimension of Forest Certification*, eds, Errol Meidinger, Christopher Elliott and Gehard Oesten. Remagen-Oberwinter: Forstbuch, 2003, pp. 63–82.

Tibor, Tom and Ira Feldman, *ISO 14000: A Guide to the New Environmental Management Standards*. Chicago, London and Singapore: Irwin Professional Publishing, 1996.

Tollefson, Chris, Fred Gale and David Haley, *Setting the Standard: Certification, Governance and the Forest Stewardship Council*. Vancouver: UBC Press, 2008.

Tomsic, Mateus and Urban Vehovar, 'Kakovost Vladanja v Starih in Novih Clanicah Evropske Unije' (The Quality of Governance in Old and New EU Member-States), *Teorija in Praksa*, 43(3–4) (2006), pp. 386–405.

UN, *Agenda 21: Programme of Action for Sustainable Development, Rio Declaration on Environment and Development, Statement of Forest Principles*. New York: United Nations Publications Department of Public Information, 1993.

————, 'Baseline Information Relevant to Specific Criteria for the Review of the Effectiveness of The International Arrangement on Forests', UN Document E/2001/42/Rev.1-E/CN.18/2001/3/Rev.1, 10 August 2004, http://www.un.org/esa/forests/reports-unff5baseline.html, accessed 07 March 2010.

————, 'Collaborative Partnership on Forests Framework 2003', UN Document E/CN.18/2003/INF/1, 20 March 2003.

————, 'Discussion Paper Contributed by the Farmers and Small Forest Landowners Major Group', UN Document /CN.18/2009/13/Add.5, 26 January 2009.

————, 'Discussion Paper Contributed by the Non-governmental Organizations and Indigenous Peoples Major Group', UN Document E/CN.18/2009/13/Add.3, 26 January 2009.

————, 'Eight-Country Initiative, Shaping the Programme of Work for the United Nations Forum on Forests (UNFF) International Expert Consultation 27 November–1 December 2000 Report of the Expert Consultation', UN Document E/CN.18/2001/2 (December 2005).

————, 'Multi-stakeholder Dialogue Note by the Secretariat. Addendum. Discussion Paper Contributed by the Business and Industry Major Group', UN Document E/CN.18/2005/3/Add.1, 24 March 2005.

————, 'Multi-stakeholder Dialogue Note by the Secretariat. Addendum. Discussion Paper Contributed by the Non-governmental Organizations Major Group', E/CN.18/2005/3/Add.4, 24 March 2005.

————, 'Multi-stakeholder Dialogue on Sustainable Forest Management', UN Document E/CN.18/2002/10, 21 December 2001.

————, 'Non-legally Binding Instrument on All Types of Forests', UN Document A/c.2/62/L.5, 22 October 2007.

————, 'United Nations Forum on Forests Bureau of the Sixth Session (UNFF 6 Bureau)' minutes of the fourth meeting, 25–26 January 2006, p. 2.

————, 'United Nations Forum on Forests', http://www.un.org/esa/forests/ accessed 03 November 2008.

——, 'United Nations Forum on Forests Report of the Fifth Session', UN Document E/2005/42-E/CN.18/2005/18, 14 May 2004 and 16–27 May 2005.

——, 'United Nations Forum on Forests Report on the Fourth Session', UN Document E/2004/42 E/CN.18/2004/17, 6 June 2003 and 3–14 May 2004.

——, 'United Nations Forum on Forests Report on the Organizational and First Sessions', UN Document E/2001/42/Rev.1-E/CN.18/2001/3/Rev.1, 12 and 16 February and 11–22 June 2001.

——, 'United Nations Forum on Forests Report on the Second Session', UN Document E/2002/42-E/CN.18/2002/42, 22 June 2001 and 4–15 March 2002.

——, 'United Nations Forum on Forests Report of the Seventh Session', UN Document E/2007/42 E/CN.18/2007/8, 24 February 2006 and 16–27 April 2007.

——, 'United Nations Forum on Forests Report of the Sixth Session', UN Document E/2006/42-E/CN.18/2006/18, 27 May 2005 and 13–24 February 2006.

——, 'United Nations Forum on Forests Report on the Third Session', UN Document E/2003/42-E/CN.18/2003/13, 15 March 2002 and 26 May–6 June 2003.

UNFF Secretariat, 'Implementation of Proposals for Action Agreed by Inter-governmental Panel on Forests and by Intergovernmental Forum on Forests (IPF/IFF): Action for Sustainable Forest Management', http://www.un.org/esa/forests/pdf/publications/proposals-for-action.pdf accessed 07 March 2010.

——, 'Review of the Effectiveness of the International Arrangement on Forests Analytical Study', undated.

UN-REDD, 'The UN-REDD Programme', http://www.un-redd.org/UNREDD Programme/tabid/583/language/en-US/Default.aspx, accessed 23 March 2010.

Utting, Peter and José Carlos Marques, 'Introduction: The Intellectual Crisis of CSR', in *Corporate Social Responsibility and Regulatory Governance*, eds, Peter Utting and José Carlos Marques. Basingstoke: Palgrave Macmillan, 2010, pp. 1–25.

——, 'Regulation in Global Governance', in *Authority in the Global Economy*, eds, Volker Rittberger and Martin Nettesheim. Basingstoke: Palgrave Macmillan, 2008, pp. 241–275.

Van Kersbergen, Kees and Frans Van Waarden, ' "Governance" as a Bridge Between Disciplines: Cross-disciplinary Inspiration Regarding Shifts in Governance and Problems of Governability, Accountability and Legitimacy', *European Journal of Political Research* 43 (2004), pp. 143–171.

Vallejo, Nancy and Pierre Hauselmann, *Governance and Multi-stakeholder Processes*. Winnipeg: International Institute for Sustainable Development, 2004.

——, *PEFC: An Analysis*. Pully: WWF, 2001.

Van Vliet, Martijn, 'Environmental Regulation of Business: Options and Constraints for Communicative Governance', in *Modern Governance: New Government Society Interactions*, ed., Jan Kooiman. London: Sage, 1993, pp. 105–118.

Vereinigung Deutscher Gewässerschutz, 'Über Uns', http://www.vdg-online.de/ueber_uns.html, accessed 19 June 2007.

Verolme, Hans J.H., William E. Mankin, Saskia Ozinga and Sophia Ryder, *'Keeping the Promise? A Review by NGOs and IPOs of the Implementation of the UN Inter-governmental Panel on Forests "Proposals for Actions" ' in Select Countries*.

Washington: Biodiversity Action Network – Tides Project/Global Forest Policy Project, 2000.

Vogt, Kristina A., Bruce C. Larson, John C. Gordon, Daniel J. Vogt and Anna Fanzeres, A., *Forest Certification: Roots, Issues, Challenges and Benefits*. Boca Raton: CRC Press, 2000.

Warren, Michael E., 'What Can Democratic Participation Mean Today?' *Political Theory* 30(5) (2002), pp. 677–701.

Wettestad, Jørgen, 'Designing Effective Environmental Regimes: The Conditional Keys', *Global Governance* 7(3) (2001), pp. 317–341.

World Rainforest Movement, *Certifying the Uncertifiable: FSC Certification of Tree Plantations in Thailand and Brazil*, ed., Larry Lohmann. Montevideo and Moreton-in-Marsh: World Rainforest Movement, 2003.

———, *FSC: Unsustainable Certification of Large Scale Tree Plantations*. Montevideo and Moreton-in-Marsh: World Rainforest Movement, 2001.

Young, Iris M., *Inclusion and Democracy*. Oxford: Oxford University Press, 2000.

Young, Oran R., 'Hitting the Mark: Why are Some Environmental Agreements More Effective Than Others?', *Environment* 20 (1999), reproduced in *Making Law Work: Environmental Compliance & Sustainable Development*, eds, Durwood Zaelke, Donald Kainaru and Eva Kružíková. London: Cameron May, 2005, Vol. 1, pp. 187–200.

Young, Oran R. and Levy, Mark A., 'The Effectiveness of International Environmental regimes', in *The Effectiveness of International Environmental Regimes: Causal Connections and Behavioural Mechanisms*, ed., Young, O. R. Cambridge MA: MIT Press, 1999.

Yrjö-Koskinen, Eero, Matti Liimatainen and Lotta Ruokanen, *Certifying Extinction? An Assessment of the Revised Standards of the Finnish Forest Certification System*. Helsinki: Greenpeace Finland, Finnish Association for Nature Conservation, Finnish Nature League, 2004.

Zaelke, Durwood, Donald Kainaru and Eva Kružíková (eds), *Making Law Work: Environmental Compliance & Sustainable Development*, vol. 1. London: Cameron May, 2005.

Zovko, Ivana, 'International Law-Making for the Environment: A Question of Effectiveness', in *International Environmental Law-making and Diplomacy Review 2005*, ed., M. Berglund. Joensuu: University of Joensuu department of Law, 2006, pp. 109–128.

Zürn, Michael and Mathias Koenig-Archibugi, 'Conclusion II: Modes and Dynamics of Global Governance', in *New Modes of Governance in the International System: Exploring Publicness, Delegation and Inclusion*, eds, Mathias Koenig-Archibugi and Michael Zürn. Basingstoke: Palgrave Macmillan, 2006, pp. 236–254.

Zyen and PEFC, 'PEFC Governance Review May, 2008 Final Report', May 2008.

Index

LEGITIMACY
204-5
5 INPUT + OUTPUT LEGITIMACY
39-40
202
158